MICROPROCESSOR SYSTEMS AND THEIR APPLICATION TO SIGNAL PROCESSING

MICROELECTRONICS AND SIGNAL PROCESSING

Series editors: **P. G. Farrell**, University of Manchester, U.K.
J. R. Forrest, University College London, U.K.

About this series:

The topic of microelectronics can no longer be treated in isolation from its prime application in the processing of all types of information-bearing signals. The relative importance of various processing functions will determine the future course of developments in microelectronics. Many signal processing concepts, from data manipulation to more mathematical operations such as correlation, convolution and Fourier transformation, are now readily realizable in microelectronic form. This new series aims to satisfy a demand for comprehensive and immediately useful volumes linking the microelectronic technology and its applications.

Key features of the series are:

● Coverage ranging from the basic semiconductor processing of microelectronic circuits to developments in microprocessor systems or VSLI architecture and the newest techniques in image and optical signal processing.

● Emphasis on technology, with a blend of theory and practice intended for a wide readership.

● Exposition of the fundamental theme of signal processing; namely, any aspect of what happens to an electronic (or acoustic or optical) signal between the basic sensor which gathers it and the final output interface to the user.

1. *Microprocessor Systems and their Application to Signal Processing:* C. K. YUEN, K. G. BEAUCHAMP, G. P. S. ROBINSON

MICROPROCESSOR SYSTEMS AND THEIR APPLICATION TO SIGNAL PROCESSING

C. K. Yuen

Centre of Computer Studies
University of Hong Kong

**K. G. Beauchamp and
G. P. S. Robinson**

Computer Services Department
University of Lancaster, UK

1982
Academic Press

A Subsidiary of Harcourt Brace Jovanovich, Publishers
LONDON · NEW YORK
PARIS · SAN DIEGO · SAN FRANCISCO · SAO PAULO
SYDNEY · TOKYO · TORONTO

ACADEMIC PRESS INC. (LONDON) LTD
24/28 Oval Road,
London NW1 7DX

United States Edition published by
ACADEMIC PRESS INC.
111 Fifth Avenue,
New York, New York 10003

British Library Cataloguing in Publication Data

Yuen, C. K.
    Microprocessor systems and their application to
    signal processing.—(Microelectronics and
    signal processing)
1. Signal processing—Digital techniques
    I. Title   II. Beauchamp, K. G.
    III. Robinson, G. P. S   IV. Series
    621.38′043′02854044   TK5102.5

ISBN 0-12-774950-0

Typeset by Preface Ltd, Salisbury, Wiltshire
and printed in Great Britain by
Thomson Litho, East Kilbride, Scotland

Preface

The present book is aimed primarily at scientists and engineers who wish to acquire a basic knowledge of microprocessor hardware and software, in order to apply microelectronics in the design and development of instruments and systems related to the measurement, acquisition and analysis of signals.

In the first part of the book we provide a fairly general introduction to the various aspects of microprocessor systems technology, including the electronics of digital hardware and its logical design, microprocessor instruction sets, input/output handling, and the use of software and hardware development tools. The second part illustrates the earlier material by showing the use of microprocessors and related devices in signal processing and data acquisition. It should be noted that as the material in the first part has a fairly general orientation, it should be useful as an aid to system development in other fields. We have not assumed prior knowledge of either microprocessors or signal processing applications in writing the first part. However, some familiarity with computing and programming will be helpful. In contrast, the second part assumes that the reader is already acquainted with various aspects of computer signal processing and data acquisition, such as Fourier transformation, correlation and filtering, and only wishes to learn the part microprocessor sytems may be able to play. Readers without such knowledge should consult some of the textbooks on signal processing and data acquisition listed in the reference sections of the book. While many readers may wish to apply microprocessors in areas other than signal handling, it is nevertheless suggested that the examples we include here provide very useful pointers to the potential of microprocessors and well illustrate their value as a system tool.

There are six chapters in Part 1 of the book. Chapter 1 gives a general introduction to microprocessor systems, including their hardware and software structures and a sketch of microprocessor systems development. It will be realized from this chapter that while microcomputer hardware systems are simpler than mainframe computers, many of the user facilities found in mainframes (e.g. operating systems and file management) have their counterparts in microprocessor systems design and that this leads to a complexity in software development for the microprocessor user.

Chapter 2 provides a brief introduction to digital electronic devices, mainly to familiarize the reader with the common terminology and the range of components employed in hardware development. Chapter 3 is concerned with the structure of common hardware units of microprocessor systems together with some brief discussion of design techniques.

Chapter 4 introduces the reader to microprocessor instruction sets and show their use in programming simple problems. Many basic ideas of computing constructs are introduced in this chapter. Chapter 5 discusses the structure of the Input/Output subsystem of microprocessors, including interrupt handling, direct memory access, interfacing, and programming considerations. Chapter 6 presents an overview of tools commonly employed in the development and testing of microprocessor software and hardware. These play an important role in making microprocessor systems accessible to the non-expert user.

Part 2 of the book is concerned with applications. Chapters 7 and 8 consider the overall aspects of microprocessor applications in signal processing and data acquisition. Components and systems which contain, or are used with, microprocessors and which perform various signal handling functions are discussed and their important properties noted. In Chapter 9 examples are given of complete microprocessor systems used to solve particular signal processing applications.

The authors take great pleasure in acknowledging the assistance of the following individuals towards this book: Professor Arthur Sale of the University of Tasmania, Dr Garth Wolfendale of CSIRO, Canberra, Dr John Hargreaves of the University of Lancaster, Terry Kennair of the University of Newcastle upon Tyne, and to the following organisations; Intel Corp., Advanced Micro Devices Ltd., TRW Products Inc., Base Ten Systems Ltd., National Semiconductor Corp., Mostek Corp., Analog Devices BV, and Plessey Ltd.

Hongkong &   *C. K. Yuen*
Lancaster, Spring   *K. G. Beauchamp*
1982   *G. P. S. Robinson*

Contents

Preface   v

PART 1   MICROPROCESSOR SYSTEMS

1   Introduction to Microprocessor Systems   3

1.1   Introduction to Microprocessor Applications   3
1.2   The Structure of Microprocessor Systems: Hardware   8
   1.2.1   Memory   9
   1.2.2   Processor   12
   1.2.3   Input/Output devices   15
   1.2.4   System configuration   22
1.3   The Structure of Microprocessor Systems: Software   24
   1.3.1   Device handlers   24
   1.3.2   Processor management   25
   1.3.3   Memory management   26
   1.3.4   Device and data management   30
   1.3.5   Error handling routines and utilities   32
   1.3.6   Application programs   32
1.4   Outline of Microprocessor Systems Development   34
   References   39

2   Digital Electronic Devices   41

2.1   Logic Gates   41
2.2   Boolean Algebra   47
2.3   Simple Logic Devices   52
2.4   Logic Circuits from NAND/NOR Gates   58
2.5   Simple Sequential Circuits   63
2.6   Logic Families   68
   References   71

3   The Logical Design of Boolean Devices   72

3.1   Synthesis and Simplification Procedures   72
3.2   Examples of Boolean Arithmetic Modules   76

	3.2.1	Decimal adder	76
	3.2.2	Carry look-ahead adder	78
	3.2.3	Comparators	81
3.3	Examples of Boolean Sequential Modules		84
	3.3.1	Registers	84
	3.3.2	Counters	87
	3.3.3	Sequential multiplier	89
	3.3.4	Floating point adder	91
3.4	Large-Scale Integration Devices		93
	3.4.1	Large memory modules	93
	3.4.2	Shift register stores	96
	3.4.3	Parallel vector processing devices	97
	3.4.4	Microprocessors	100
	References		102

4   Microprocessor Programming   **103**

4.1	Overview of Microprocessors	103
4.2	Introducing the Intel 8080 Microprocessor	109
	4.2.1	Program counter	111
	4.2.2	Branches and condition codes	118
	4.2.3	Stack pointer	121
4.3	Intel 8080 Machine Instructions	125
	4.3.1	Data movement instructions	127
	4.3.2	Jump instructions	129
	4.3.3	Accumulator manipulation instructions	130
	4.3.4	Register manipulations	135
	4.3.5	Miscellaneous	136
4.4	Assembly Programming	136
	4.4.1	Pseudo-codes and two-pass assemblers	137
	4.4.2	Macros	140
	4.4.3	Use of subroutines	142
	4.4.4	Final example	144
4.5	The Software Development Process	147
4.6	Directions for Microprocessor Architecture Development	149
	References	152

5   Microprocessor Input/Output Handling   **153**

5.1	Simple I/O Systems	153
	5.1.1	Basic I/O hardware	155
	5.1.2	Simple I/O programming	161
	5.1.3	Intel 8085 and other manufacturers' I/O systems	164
5.2	Interrupt Systems	166
	5.2.1	Intel 8080 interrupt handling	169
	5.2.2	Interrupt control hardware	172
	5.2.3	Interrupt programming	178
5.3	Direct Memory Access Systems	183
5.4	Interfacing Standards	186
	5.4.1	RS232 interface	187
	5.4.2	IEEE–488 bus	187

5.4.3 Intel multibus 190
References 193

6 Development Systems 194

6.1 Software Development Aids 195
    6.1.1 The process of software development 195
    6.1.2 Assemblers 196
    6.1.3 High-level languages 198
    6.1.4 Simulators 201
    6.1.5 Microprocessor operating systems 202
6.2 Hardware Development Aids 204
    6.2.1 The process of hardware development 204
    6.2.2 Hardware design aids 205
    6.2.3 Hardware testing and debugging aids 207
References 211

PART 2 SIGNAL PROCESSING APPLICATIONS

7 Microprocessors in Signal Processing 215

7.1 Introduction 215
7.2 Microprocessor Signal Handling 216
7.3 Data Logging 218
7.4 Transformation 223
    7.4.1 The fast Fourier transform algorithm 224
    7.4.2 FFT processors 230
7.5 Correlation 234
    7.5.1 Correlation using the FFT 235
    7.5.2 Bit-by-bit correlation 237
    7.5.3 Polarity coincidence correlation 239
7.6 The Programmable Signal Processor 241
    7.6.1 Hardware design 242
    7.6.2 Software for the PSP 246
References 247

8 Analog/Digital Operations 249

8.1 Introduction 249
8.2 Analog/Digital Conversion 249
    8.2.1 The operational amplifier 250
    8.2.2 The digital/analog converter 253
    8.2.3 The analog/digital converter 255
    8.2.4 Controlling the conversion process 259
8.3 Analog I/O Connection 262
8.4 Digital Filtering 266
    8.4.1 Design considerations 268
    8.4.2 The IIR filter 269
    8.4.3 The FIR filter 278
References 281

9 Applications   282

9.1 Acoustic Imaging   282
    9.1.1   Historical perspective   282
    9.1.2   Phased arrays   286
    9.1.3   Acoustic holography   287
    9.1.4   An acoustic imaging system   290
    9.1.5   The phased array transmitter   292
    9.1.6   The holographic receiver   293
    9.1.7   System control   297
9.2 Data Logging of Solar Activity   297
    9.2.1   Measurement of solar activity   299
    9.2.2   A microprocessor-based riometer unit   300
    9.2.3   The replay unit   306
    9.2.4   Summary   306
    References   307

Index   308

Part One

Microprocessor Systems

Introduction to Microprocessor Systems

1.1 INTRODUCTION TO MICROPROCESSOR APPLICATIONS

The microprocessor represents one of the more recent developments in digital electronics and integrated circuit technology. First appearing in the early seventies in the form of crude devices with limited instruction sets and interfacing capabilities, by the end of the decade the microprocessor had become a highly complex and sophisticated unit with capabilities rivalling minicomputers and even large mainframes. Microprocessors are now employed in a vast array of applications. Simple processors are found in household appliances to implement intelligent and flexible machine control, while the larger and newer processors appear increasingly in personal and business computing systems. Few of today's machines and instruments are without a microprocessor somewhere within the whole unit, and even more widespread application may be expected as microprocessors continue to become cheaper, more powerful and easier to use.

However, by itself, a microprocessor is a device of very restricted capabilities. It can do little more than accept a set of numbers (*input*) perform some arithmetic or other manipulative operations on them (*processing*), and either temporarily hold the results in its internal storage (*registers*) or *output* the results to the other parts of the microprocessor system. Moreover, it can only handle data of a very special form, namely *binary* data—numbers made up of strings of ON/OFF electrical pulses (ones and zeros). Thus, before information produced by humans or instruments may be processed by a microprocessor, it must first be converted into binary form by the *input/output* (I/O) subsystem attached to the microprocessor, and stored as strings of zeros and ones in the *memory subsystem*. Appropriate interconnection (*interfaces*) must be established between the processor, the memory and the I/O units, such that the processor may cause the I/O units to acquire the needed data and place them into suitable locations in the memory. We shall consider various aspects of the hardware of microprocessor systems in the following section and in Chapters 2 and 3.

It may also be said that the *kind* of data manipulations a microprocessor can perform is also highly restricted—such as copying a number from one storage location to another, adding or subtracting two numbers, or testing whether a number is positive, zero or negative, or larger or smaller than another. Before a new microprocessor is produced, its designers must first decide on a list of operations the processor should be able to perform, and this list is called the processor's *instruction set*. The details of the processor's internal circuits are then worked out to make it possible to perform such operations (or in computing terms, to *execute* such instructions).

We may ask: What accounts for the power and versatility of microprocessors? The answer lies in their great speed. They can accept, process and return hundreds of thousands or even millions of items of data per second. By performing many individually simple operations in rapid succession and in a coherent fashion, a microprocessor is able to accomplish many complex tasks, even those which apparently have little to do with arithmetic or number testing.

Take, for example, the control of a washing machine. The following is a possible sequence of operations performed by the machine controller (which may be a microprocessor, a mechanical controller, or even a person operating knobs and switches) after the machine is turned on.

1. Reset all parts of machine to idle.
2. Turn on tap.
3. Test water level indicator. If "full" go on to step 4. Otherwise repeat step 3.
4. Turn off tap. Turn to "wash".
5. Wait 10 minutes.
6. Turn off wash. Turn on pump to empty.
7. Test water level indicator. If empty go on to step 8. Otherwise repeat step 7.
8. Turn to "spin".
9. Wait three minutes.
10. Turn off spin . . .

We may see that some of the operations involved are input operations, which bring data (e.g. water level) about the state of the machine into the control unit for analysis. Some are tests of data values (water full, empty, or neither full nor empty). Yet others are output operations, e.g. sending an On or Off pulse to the motor switch. Arithmetic may also be required, such as step 5, which requires the controller to keep count of elapsed time.

The above illustrates the idea of *programming*. As explained earlier, a computer processor has a number of in-built functions that it can be made to perform, which make up its instruction set. Programming is the process of specifying a sequence of instructions which would together perform a particular task. A computer program is just a set of logically related instructions. It is written in a computer *language*, which is a system of notations for defining the program, by specifying what operations are required on what data. As each individual microprocessor model has its own set of operations, it has to have its own machine language, though there is a great deal of similarity between the languages of different processors. Also, each problem may be programmed in many languages. A *high-level* or *problem-oriented* language provides a notation particularly related to a class of problems and therefore gives concise and comprehensible descriptions of their solutions. The machine language of a normal microprocessor, on the other hand, is not designed to suit particular applications, but relates directly to the hardware design. In consequence, specifying the solution of a problem in machine language instructions is usually a complex and tedious task. Even a simple task like washing machine control requires a program containing many instructions, all of which must be correctly specified to make the program work. As we shall discuss later, there are programming tools which help to reduce the difficulty of software development. In particular, it is possible to specify the solution of one's problem in a high-level language and then use a computer to translate the program into the machine language for one's microprocessor.

Regardless of the actual software development process involved, the final product is always a machine language program consisting of strings of ones and zeros, which will be accepted by the microprocessor and cause it to perform the set of operations specified by the program. Because of the large number of instructions involved, a microprocessor program cannot be stored in the microprocessor itself, but is kept in the memory subsystem, which also makes space available for data. The program will co-exist with data of various types including constants, reference tables, data received from input devices but as yet unprocessed, results returned from the processor but not yet sent to output devices, and intermediary results which will be used in later processing and then erased. Programs and the various types of data may reside in the same memory modules or separately depending on conditions (see next section).

A memory module comprises semiconductor storage "cells" (whose structure we shall study in Chapter 2), each of which can have two possible states, 0 or 1 (ON or OFF). The module also contains some control mechanism which can select a particular group of cells and reproduce the cells' contents as electrical pulses, which may then be sent to the microprocessor as data to be worked on. This operation is called a memory READ. The same control mechanism can also receive electrical pulses and store their binary values in a designated part of the memory, and this is called a memory WRITE. Each individual cell is said to contain one *bit* of data.

During the operation of a microprocessor system, the processor *executes* a program by reading the instructions in the program from memory one by one, analysing each instruction to see what operation is to be performed on which data, and then fetches the data from the memory, manipulates them as required, and returns the new data to the memory, all as specified by the instruction. The first step is called an instruction *fetch*, the second step an instruction *decode*, and the last step the actual *execution* of the instruction. Most instruction executions involve passing data from and to the memory, some affect only data already inside the processor, while others, the I/O instructions, pass data to or from I/O devices, e.g. reading the setting of a switch or sending a string of numbers to a printing wheel. As we shall see later, special programming techniques are required for the purpose of controlling I/O devices and effecting data transfers between them and other parts of the microprocessor system. The interconnection between the various parts of the system is also a fairly complex task if proper data and control signal transmission is to be ensured.

We have identified three main components of a microprocessor system: The processor (or processors, since a system may well have multiple processing units sharing common memory and I/O devices), one or more memory modules with a control mechanism making up the memory subsystem, and the I/O devices with their interfacing hardware making up the I/O subsystem. Generally speaking, as far as hardware cost is concerned the I/O subsystem constitutes the largest expenditure. Whereas processors and memory modules handle only digital data in the form of binary electrical pulses of specific values and durations, I/O devices function to convert data between binary digital form and other forms, such as manually controlled switches, printed or other visually displayed text, analog (i.e. continuously variable) electric voltages or currents, magnetically recorded information, temperature, pressure, etc., and thus contain a wider variety of circuit elements. The present semiconductor technology allows us to implement circuits of great complexity on a small piece of silicon and produce elaborate microprocessors and memory modules at low cost.

They may be mass produced because the same microprocessor and memory components may be applied to different tasks by different programming—a factor that makes processors and memory modules cheap. In contrast, I/O devices usually contain mechanical or other non-electrical mechanisms, and there is great variation between devices in the type of information they handle, the data rates involved, and the control mechanism. This makes component fabrication more difficult, especially as some components have to handle fairly large currents to drive the I/O mechanisms. Further, as each microprocessor system has its particular configuration of devices to suit the problem it is to solve, the I/O subsystem needs to be specially put together for it, and one is less able to take advantage of mass production methods.

However, by far the greatest individual cost item in the development of a microprocessor system is the program or software. Not only is the final running program itself a highly complex product (even for relatively simple applications like washing machine control), and therefore takes much expensive skilled labour to design, test and modify before it may be made to work, but a great many costly facilities will have to be made available to make the software development process possible. Indeed, it often requires a computer system to develop programs for a microprocessor system, as we shall discuss in Chapter 6.

We now begin to see that, whereas microprocessors are cheap, a working microprocessor *system* seldom is. We also see that the economic advantage of employing a microprocessor system for any particular application is dependent on the following factors.

1. The application should preferably be one that occurs widely in the same form or several fairly standard forms, so that a single system configuration with a standard program may be implemented in many replicated units and the cost of the hardware design and software development may be divided among many individual units, resulting in a low development cost overhead per unit.

2. The application should involve simple I/O devices of standard types and have standard data rates. The processing involved should be reasonably simple, so that it is within the capability of standard, mass produced processors executing relatively simple user-developed software together with standard system software.

While microprocessor systems can be, and indeed are, widely applied to problems which do not meet these requirements, the development cost involved is likely to be considerably higher.

The economic factors are largely responsible for the high degree of user involvement in the hardware and system software aspects of microprocessor systems. Whereas users of large computer systems usually need just to know how to write their programs, with microprocessor systems the user often has to design and implement hardware and system software himself, because it would be far too expensive to have the work done for him by others. Again, whereas large computer systems always have an elaborate system of manufacturer support for both hardware and software, it is not economical for microprocessor producers to provide the same degree of support. Thus, not only has the development of microprocessors had a major impact on the application areas (including our domestic life, through such things as microprocessor-controlled appliances), it has also affected the computing profession, by once again requiring it to train people in all aspects of computer systems rather than just as narrow specialists. For this reason also, the present book will introduce the reader to both the hardware and the software aspects of microprocessor systems development.

1.2 THE STRUCTURE OF MICROPROCESSOR SYSTEMS: HARDWARE

In the present section we give the reader an overview of the hardware structure of microprocessor systems, discussing the basic components of these systems as well as the interconnection of these to form a system configuration. The discussion is general and centres on the "functional" aspects, rather than details of circuit design or construction. However, first we must make some remarks on current digital hardware technology.

Virtually all the electronic parts of current microprocessor systems are fabricated by *integrated circuit* (IC) techniques, in which a large number of electronic components (transistors, together with diodes, resistors, capacitors, etc.) are constructed and interconnected on a single piece of silicon. The resulting IC "chip" can have very great internal complexity, e.g. 2^{16} memory cells or a complete microprocessor contained on a piece of silicon of about one centimetre square. Although these techniques enable continually increasing circuit complexity to be contained on a single chip, they do give rise to a "switching problem". It is only possibly to connect so many wires to a small piece of silicon, the maximum at present being about 64. Thus, information to be handled by the IC devices, and results produced from them, all have to pass over these small number of connections, so that, despite the high speed and internal elaboration of the devices, there are limitations in our ability to take advantage of them in data handling. In fact, a great proportion of the circuits are included in each IC chip to direct information to and from its complex internal structure through the small number of connections. The various individual IC components of a microprocessor system have to be designed to operate together with only relative small numbers of interconnections between them.

This has several implications. First, data transfers take place in units of small size, such as 4, 8, 16, or occasionally, 32 bits. Further, it is insufficient to transfer data by themselves. There has to be accompanying information about which component as well as *where within the component* the data have to be directed to or accessed from precisely because each component may have considerable internal complexity. Also, it is necessary to specify what action the receiving device is to perform on the data. We see that a data transfer in a computer system has to be in a "package" containing three different types of information; a *unit of data, addressing information* (where), and *control information* (what to do). And there are different levels of elaboration in the way the "packages" are conveyed. On larger systems the different parts may travel simultaneously over connections provided separately, while on smaller systems they may travel successively over the same set of wires. Obviously, the latter arrangement reduces system speed, since each data transfer operation takes several times as long. Consequently, the "bus width", i.e. the total number of separately provided electronic pulse paths, has a major effect on overall system speed. On the most sophisticated systems a large bus width is provided, but data may be passed in units of several sizes, which requires yet more control information to indicate how long the data item is.

A second implication of restricted chip connection is that data processing has to be predominantly serial even where parallel data manipulation hardware could have been provided at competitive cost. Take, for example, the addition of two vectors. Although we can fabricate IC chips capable of adding simultaneously, say, 256 pairs of numbers, few microprocessor systems take advantage of such devices because we cannot transfer 256 pairs of values simultaneously between the data stores and the parallel addition device. Instead, values have to be sent a few at a time, and the high speed potentially available cannot be utilized. The result is little better that the serial method of the processor fetching a pair of numbers at a time, adding them, and returning the sum before proceeding to fetch the next pair.

We are now ready to discuss individual system components, and commence with the memory.

1.2.1 Memory

As remarked earlier, it is now common to have IC chips containing 2^{16} individual memory cells, grouped into units of 8 or 16 bits. Each unit is termed a memory *word* or a memory *location*, and its length is the *word-length* of the memory. An 8-bit unit is given the special name of a *byte*. Processors and computers that manipulate data in 8-bit units are called byte machines. It is, of course, possible for a machine with a particular word-length to have also the capability of handling data of other sizes. A common size for memory modules is 64K or 2^{16} bits, although 256K modules are becoming readily available.

A 2^{16}-bit memory chip with word-length 8 would have 2^{13} individual memory locations. Each of these is identified by an *address*, from 0 to $2^{13} - 1$, or in binary form, 0000000000000 to 1111111111111. Thus, the chip would have a maximum of 13 pins or connectors for addressing and 8 connectors for the data. Figure 1.1 shows the pin connections for the Intel 8316A Memory Chip. It will be seen that 3 connectors are allocated for select pulses. These are required to indicate to the memory control circuits whether one wishes to place data in a memory location or take data out of it, i.e. to carry out a READ or a WRITE operation. To WRITE into a memory location it is necessary to place the data digits on the data lines of the memory module, place the desired address of the memory location on the address lines, and then send a pulse along the WRITE control line. This will cause the data direction circuits to accept the data from the data lines and place their values into the specified memory location, erasing the values previously stored there. To READ from a memory location it is only necessary to place the address on the address lines and send a pulse along the READ control line. This will cause the data direction circuits to copy the bit values stored in the location addressed onto the data lines, ready for transmission elsewhere in the system. The READ operation does not change the values stored in a memory location, and the same contents can be read as many times as needed during various stages of processing.

PIN CONFIGURATION

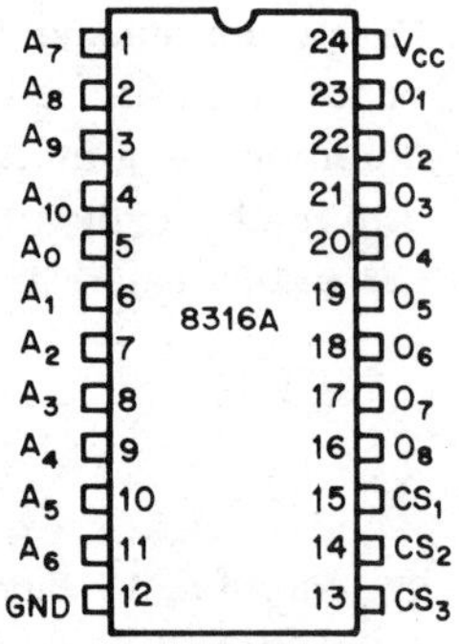

```
         A₇ ⊏ 1        24 ⊐ V_CC
         A₈ ⊏ 2        23 ⊐ O₁
         A₉ ⊏ 3        22 ⊐ O₂
        A₁₀ ⊏ 4        21 ⊐ O₃
         A₀ ⊏ 5        20 ⊐ O₄
         A₁ ⊏ 6  8316A 19 ⊐ O₅
         A₂ ⊏ 7        18 ⊐ O₆
         A₃ ⊏ 8        17 ⊐ O₇
         A₄ ⊏ 9        16 ⊐ O₈
         A₅ ⊏ 10       15 ⊐ CS₁
         A₆ ⊏ 11       14 ⊐ CS₂
        GND ⊏ 12       13 ⊐ CS₃
```

PIN NAMES

$A_0 \cdot A_{10}$	Address inputs
$O_1 \cdot O_8$	Data outputs
$CS_1 \cdot CS_3$	Programmable chip select inputs

Fig. 1.1. Pin assignments of a 16K memory chip.

It is clear from the discussion that a memory module contains two functionally distinct parts; the memory cells, and the addressing circuits which can connect selectively some of the cells with the data lines to permit READ or WRITE operations to occur. But both types of circuits are made from the same basic components, namely transistor switching devices. We shall discuss in Chapter 3 the principles of their construction.

In the foregoing we have used the term "data" in a rather general sense. Computer memories contain several kinds of data. Besides numerical or text data being processed, the same memory modules are also used to store the computer program that processes the data. In order to obtain the data, the stored program has to know their addresses, some of which may be contained within parts of the program, and others may be stored in separate memory locations and, as far as the program is concerned, are treated just like data. Generally we can identify at least three types of data stored in the memory; genuine data, addresses showing where to find the data, and programs that use the addresses to get the data. As far as memory modules are concerned, however, one kind of data looks just like another; they are all strings of ones and zeros. The distinction between them is a logical one; it is up to the programmer to specify where the different parts are and to use them appropriately. This does not mean, however, that the programmer has to remember exactly the address of every instruction in his program and every item of data at all times, since the software development tools will do much of the work involved. It may also be seen that not all the memory READ instructions would result in data being accessed, since addresses showing where to find them may first have to be fetched from memory. Sometimes addresses may have to be produced by computation using other stored values, and there may be quite elaborate juxtaposition of program/address/data accesses. We shall study these various possibilities in Chapter 4.

A microprocessor system may contain several different types of memory modules because of the varying patterns of memory accesses. Some data items are accessed frequently, and it is desirable to place them in memory modules designed for high-speed access. Others may be stored in slower and cheaper modules without appreciable loss of overall processing speed. Often, parts of the data—in particular programmes and tables—do not change during processing, and indeed must be protected against accidental alterations. These are usually placed in *read-only memory* (ROM) chips, which cannot change their contents through normal WRITE operations because of their special fabrication. The contents of the ROM modules may be set at the time of chip manufacture or by the purchaser using special equipment to insert his data into a *programmable* ROM or PROM. Some PROM chips can never change their contents once set, but an alternative *erasable* PROM is available whose contents may be erased by ultraviolet radiation and then re-set using the normal PROM programming equipment. These are known as EPROM memories. Yet others may be re-programmed simply by manipulating their control inputs in a specified way, even though the contents may not be altered by normal memory WRITE instructions.

Most current microprocessors require more than 2^{16} bits of memory for program and data storage. In consequence, such a system will contain several memory chips, possibly of more than one type. To distinguish between the memory location of one chip from those of another, the system requires additional address lines, besides those that connect to the address lines of memory modules, in order to indicate which module is to be accessed. They also require additional addressing circuits which will direct READ and WRITE control pulses to the memory chip according to the 0/1 values contained in the additional address lines. Again, we shall see later how circuits for this purpose may be constructed. In a large-scale microprocessor system there are additional memory control and management functions, to be provided by yet other hardware components. Discussion of these will be deferred to later sections.

1.2.2 Processor

Like memory modules, microprocessors are circuits of high internal complexity fabricated onto small pieces of of silicon, with the usual restriction on the number of pins available. Much of the processing capability of a unit is therefore devoted to moving data between the memory and the circuits in the processor that actually carry out operations on the stored data.

In order to obtain data from particular memory locations, the processor must place their addresses on the address lines of the system. A number of the available connectors of a microprocessor are therefore address pins, for connection to the address lines. There are also internal circuits which will generate digital pulses according to the value of the memory address wanted, and to place them on the address lines, and circuits to generate READ or WRITE pulses for the memory control lines. Additional pins are available for other control purposes, such as power supply, ground and microprocessor START line, those related to input/output operations and data transfer controls, together with various *status* lines for conveying information about the condition of I/O devices and the system in general. Inside the microprocessor, circuits are provided to transfer pulses to the control lines, or to accept pulses from the status lines. These parts comprise the external interface components of the microprocessor. Figure 1.2 illustrates the various paths of a typical microprocessor system.

Three other main components may be identified. The first of these are the data processing circuits which manipulate the data received into the proces-

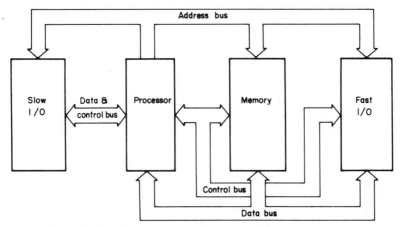

Fig. 1.2. Configuration of typical microprocessor system.

sor, performing such operations as addition, subtraction, logical operations, testing a data item for value and returning information about it (whether it is zero/nonzero, positive/negative, larger/smaller than another data item, etc.).

Secondly, we have data storage circuits for holding a small number of data items which are needed frequently during the execution of a program so that they need not be fetched repeatedly from the memory. This comprises a number of high-speed memory cells, grouped into words of standard sizes. Each group is called a *register*. Obviously a processor with more registers is likely to perform better because it can keep more data in rapidly accessible places. Unfortunately, we shall see later that this causes a problem with instruction formats and in consequence most microprocessors have only a small number of registers, usually around 8 or 16.

Finally there is the execution control component, which interprets the instructions and passes control information to the appropriate components. As we mentioned in Section 1.1, a microprocessor program consists of a stream of instructions specifying step by step the data processing operations required by the programmer. It is the execution control unit which causes the processor to fetch the instructions of a program in sequence and to analyse each of them to determine what needs to be done before generating control pulses to the other components of the processor to effect the required operation.

To illustrate this process consider the execution of a program which has been previously placed into memory (and we shall discuss later how this is done), starting at memory location *A*. This is illustrated in Fig. 1.3. Address *A* is first loaded into the execution control unit of the microprocessor (and we shall consider later how this occurs), and the START signal is initiated. The

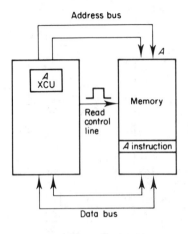
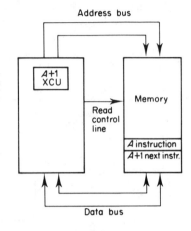

(a) Send address A to address line of   (b) Instruction read out from location A
    memory. Then send READ pulse   is sent to processor. Address A in XCU
                                                increased to refer to next instruction

Fig. 1.3. Bus operations to perform instruction fetch.

microprocessor then enters the *instruction fetch* cycle, during which the execution control unit places the program address it already holds, namely the value A, onto the address lines, generates a READ pulse, and also opens a path between itself and the data lines of the memory subsystem. The READ pulse would cause the content of memory location A to appear on the data lines, and therefore enter the execution control unit. In this way, the unit obtains the first instruction of the program, which it analyses to decide what steps are required to execute it. After the execution is completed for this instruction, the control unit would increase the program address it is holding from A to $A + 1$, and re-enter the instruction fetch cycle, which would then provide the next instruction of the program to execute.

Now let us now look into the instruction execution process. Suppose the instruction just fetched turns out to be "add the values in registers 0 and 1, then store the result in register 7". The execution control unit, recognizing this instruction, would then open data paths from the registers 0 and 1 to the input lines of the ADD unit, wait for the addition to finish, and then open a path between the output lines of the ADD unit and register 7, which completes the execution.

To take a more difficult example, suppose the instruction is "add the values stored in memory locations X and Y, and store the sum back into location Y". To do this the execution control unit would first cause the address of memory location X to be placed on the address lines of the system, and generate a READ pulse. This causes the value stored in X to appear on the data lines. The control unit then connects the data lines to the input lines of the ADD unit. It then fetches the content of Y and sends this to the ADD unit in the same way, before connecting the output lines of the ADD unit to the data lines of the memory. With the sum now on the data lines, the execution control unit replaces the address of Y on the address lines, and sends a WRITE pulse, which causes the sum on the data lines to be stored in the location Y, to complete the execution.

It may be seen that much of the execution control function consists of switching operations, and the control unit is just an elaborate network of transistor switches. We shall see in the next section how networks of this type work. Further, we shall see that the arithmetic unit, memory modules and indeed most other components of a microprocessor also consist of switching networks, which can be made to serve a large range of diverse functions.

1.2.3 Input/Output Devices

In the recent subsection we provide a general discussion of the hardware of common I/O devices. Although the circuits contained in an I/O device are much less complex than those contained in a microprocessor chip or a memory module, the hardware is still fairly complex. Further, because I/O devices contain electromechanical or magnetic components, which take much greater currents to operate than memory cells or processor components, the electronic parts needed in I/O devices cannot be produced in the same minute sizes, and individual IC chip in I/O devices do not have the same internal circuit density. In consequence, in I/O devices even relatively simple functions may take several chips to achieve, the number increasing, usually, with the speed and the data capacity of the device, the procedural complexity for device control, and so on.

Functionally we can identify two components in an I/O device. First there is the mechanism which interacts with an external medium to perform the I/O operations. For example, in a teletypewriter terminal the printhead causes visible marks to appear on paper, and the keyboard accepts button pushes by a human hand. We may also include here the circuits that control the mechanisms. Again using the terminal as an example; the electronics controlling the printhead accept a sequence of pulses whose 0/1 values specify that a particular symbol is to be printed, and cause the head to make the mark before moving it forward one slot ready to print the next symbol when requested. We would also find that each depression of a keyboard button will initiate controlling circuits which generate a unique sequence of 0/1 pulses representing the character required. Additional circuits are required for such purposes as keyboard locking, upper/lower case control, local/line transmit modes, variable symbol sets. In addition, the data actually sent to or received from the microprocessor may be in "packaged" form, and circuits must be included to process such information.

The second main component is circuits for interfacing the I/O device to the microprocessor system. There are switching circuits connecting the data lines of the device to the data lines of the microprocessor system when the address/ control signals indicate that the device is to perform an I/O operation. In addition, there must be connections along which the processor can detect the status of the device, e.g. whether it is available, whether it has successfully performed the last operation and is therefore ready for the next one, whether an error condition exists (I/O devices, being electromechanical, are more prone to incorrect information transfer). Moreover, whereas the processor and the memory both operate at high transmission rates using electronic pulses of very short duration, I/O devices are slower, and require control, status and data pulses over extended periods. These are achieved by including in the interface *buffering registers* which can hold both data and control signals. Thus, the processor would write control and data values into these registers, which then hold their values throughout the period required by the I/O devices. When status information is generated in an I/O device, it too is stored in a register for access by the processor. More elaborate circuits are needed in I/O interfaces for processor–memory–I/O device interactions taking the form of interrupts and direct memory access. These will be considered later in Chapter 5.

Let us now consider a number of common I/O devices.

(a) *Terminals*   A terminal consists of two devices; an input device (keyboard) and an output device (data printing/display unit). Generally these are contained within a single unit. Those that have printheads and produce a paper output record are termed hardcopy devices, and various designs of print mechanisms are available, including typewriter "golf balls" and "daisy wheel" devices. These devices have different symbols molded around the head periphery and printing is arranged by rotating the printhead to bring the required character to face the paper and then initiating a hammer strike on the paper. An alternative printing system is the "dot matrix" mechanism in which a group of pins are activated to form symbols as a matrix pattern of fine dots. Generally speaking, golf ball and daisy wheel devices are more expensive but produce better output of better visual quality, and can provide different symbol fonts through a change of wheel or ball. Some high-quality dot matrix terminals are also available having a larger than usual number of pins, giving a finer resolution to the symbols formed. With these improved devices different fonts may be generated simply by altering the patterns of dots.

Since hardcopy terminals are expensive and require mechanical maintenance, most low-cost microprocessor systems incorporate visual display units (VDUs). These form symbols on a television type screen in a fashion similar to a dot matrix arrangement and indeed some personal computer systems simply employ ordinary domestic television sets for this purpose. Although such units are capable of greater output speed than printing devices, most of the low-cost VDUs operate at the standard rate of 30 characters per second to avoid non-standard interfacing requirements. A screen may also be used to form pictures as well as letters and numbers, when they are termed graphics terminals. They can operate at a variety of line speeds.

Printers work on the principle that each printable symbol corresponds to a unique binary number, so that sending any particular number to the printer will cause the corresponding symbol to be printed. Such a set of one-to-one equivalences is called a symbol table. There are in fact several such standard tables, the most common is the 7-bit ASCII (American Standard Code for Information Interchange) code, shown in Table 1.1. The table shows, for example, that pushing button A on the keyboard will produce the pulse train 1000001. Normally there is a mandatory eighth bit (to make up one byte of information) plus several bits for data transmission purposes. Table 1.1 shows that, of the 128 available pulse sequences only approximately 90 correspond to printable symbols. The remainder are allocated as control symbols, etc. These are referred to as "non-printing" characters. For example, sending the LF (linefeed) symbol to the printhead would not result in a printed output to appear but would cause the printhead to move down to the next line (without moving sideways). The CR (carriage return) symbol is also unprintable, and causes the head to return to the start of the current line (without moving up or down the page). When a line is complete and a new line is to be started, the microprocessor has to send to the terminal both CR and LF to make the head go down to the leftmost position ready for the next line of print.

(b) *Floppy discs*   Floppy discs provide the most common means of storing large (by microprocessor standards, that is) amounts of information in a rapidly accessible form. The information may be data to be processed, output to be transmitted (e.g. printed), intermediate results too large in number to be retained in memory modules, machine programs available for transfer into memory for execution, source programs that need to be translated before they can be executed, or programs in a state of partial development.

Information is stored as long strips of magnetic dots on a 5 in. to 8 in. diameter plastic disc, coated with magnetic material and stored inside a paper sleeve, having a hole in the centre for locating on the *disc drive* unit. The sleeve has a slit along a radius through which the READ/WRITE head of the disc drive may record or play back data on the disc. The surface of the disc is

Table 1.1 The 7-bit ASCII code.

← PRINTABLE CHARACTERS →									CONTROL CHARACTERS		
CHAR	OCTAL	BINARY	CHAR	OCTAL	BINARY	CHAR	OCTAL	BINARY	CHAR	OCTAL	BINARY
A	101	1000001	a	141	1100001	SP	040	0100000	NUL	000	0000000
B	102	1000010	b	142	1100010	!	041	0100001	SOH	001	0000001
C	103	1000011	c	143	1100011	"	042	0100010	STX	002	0000010
D	104	1000100	d	144	1100100	#	043	0100011	ETX	003	0000011
E	105	1000101	e	145	1100101	$	044	0100100	EOT	004	0000100
F	106	1000110	f	146	1100110	%	045	0100101	ENQ	005	0000101
G	107	1000111	g	147	1100111	&	046	0100110	ACK	006	0000110
H	110	1001000	h	150	1101000	'	047	0100111	BEL	007	0000111
I	111	1001001	i	151	1101001	(	050	0101000	BS	010	0001000
J	112	1001010	j	152	1101010	)	051	0101001	HT	011	0001001
K	113	1001011	k	153	1101011	*	052	0101010	LF	012	0001010
L	114	1001100	l	154	1101100	+	053	0101011	VT	013	0001011
M	115	1001101	m	155	1101101	,	054	0101100	FF	014	0001100
N	116	1001110	n	156	1101110	-	055	0101101	CR	015	0001101
O	117	1001111	o	157	1101111	.	056	0101110	SO	016	0001110

Binary	Octal	Char		Binary	Octal	Char		Binary	Octal	Char		Binary	Octal	Char
0001111	017	SI		0101111	057	/		1010000	120	P		1110000	160	p
0010000	020	DLE		0110000	060	0		1010001	121	Q		1110001	161	q
0010001	021	DC1		0110001	061	1		1010010	122	R		1110010	162	r
0010010	022	DC2		0110010	062	2		1010011	123	S		1110011	163	s
0010011	023	DC3		0110011	063	3		1010100	124	T		1110100	164	t
0010100	024	DC4		0110100	064	4		1010101	125	U		1110101	165	u
0010101	025	NAK		0110101	065	5		1010110	126	V		1110110	166	v
0010110	026	SYN		0110110	066	6		1010111	127	W		1110111	167	w
0010111	027	ETB		0110111	067	7		1011000	130	X		1111000	170	x
0011000	030	CAN		0111000	070	8		1011001	131	Y		1111001	171	y
0011001	031	EM		0111001	071	9		1011010	132	Z		1111010	172	z
0011010	032	SUB		0111010	072	:								
0011011	033	ESC		0111011	073	;								
0011100	034	FS		0111100	074	<								
0011101	035	GS		0111101	075	=								
0011110	036	RS		0111110	076	>								
0011111	037	US		0111111	077	?								
1111111	177	DEL		1000000	100	@								

divided into a number of circular tracks, and data are recorded by the READ/WRITE head by placing 0/1 magnetization dots serially along a given track. These are played back by reading the magnetic state of the dots through the magnetic read head. As a floppy disc has multiple tracks along its surface but there is only one READ/WRITE head per disc drive, a READ or WRITE operation is normally preceded by a SEEK operation to locate the head above the track containing the required data area. Since SEEK is a start/stop movement, it takes considerable time, whereas the actual READ/WRITE operation is very rapid being related to the rate of rotation of the disc. As a result, reading a single item of data is likely to take nearly as long as reading a whole track of data. In order to be time-efficient, data transforms to and from a disc are arranged in large blocks. However, too large a block size would be space-inefficient, since we may not always be able to fill every block with data, hence a fixed size of 128 or 256 bytes per block is accepted as a compromise standard.

In addition to the SEEK, a disc operation also has to wait until the start of the block required is located under the READ/WRITE head. This waiting time is called the *latency* time. Each operation takes at least the sum of three times: SEEK, latency, and READ/WRITE time. Moreover, disc operations are prone to errors because of surface imperfections, dust or non-consistent rotation. Because of this, built-in circuits are included in the disc interface and software in the disc I/O routine to detect common errors, and to repeat any failed operations until their successful completion.

Before a new disc may be used, it first has to be "formatted". That is, magnetic dots identifying the start and end of each block are placed into position, and timing dots for checking rotations speed are put on to some or all of the tracks. A disc block *directory* also needs to be laid out. The directory indicates what data currently reside in each block. Every WRITE operation requires a disc directory update, since a previously unused (free) block has now been given current data. Before each READ or WRITE operation a directory search has to be made, either to find which block contains the required information, or to locate a convenient free block to write data into. The directory also records those block locations which cannot be used because of surface defects, so that the program would not select these. Usually, when a disc is placed into a drive and made available to the microprocessor, the control software would copy the directory from the disc and place into memory, so that it may be searched and updated rapidly.

We see that the electronic circuits for controlling disc units are considerably more complex that those of other devices, such as terminals. The interface mechanism is also more complex. Because discs have higher speed and handle larger blocks, more buffering facilities have to be built in, as well as byte-counting mechanisms to ensure that READ or WRITE operations stop when the end of block is reached. The status indicators are also more elaborate, since a disc may be free, in a SEEK, in a rotation wait, in a READ/WRITE, or finished READ but waiting for data to be picked up. Further, with disc and other high-speed devices the processor has to respond to a READ/WRITE completion rapidly to avoid unproductive waiting and the possibility of data loss. We shall consider later the hardware and software provision needed to achieve this.

(c) *Analog interfaces* Microprocessor systems, especially those employed in signal processing, are likely to handle various types of non-digital devices. Signals such as voice, light, temperature or pressure, first have to be *transduced*, i.e. measured by an instrument which outputs an analog electric voltage approximately proportional to the value of the external input. The voltage is then converted into digital form, using an analog-to-digital (A/D) converter for input into the computer system. On the output side, digital-to-analog (D/A) conversion first takes place, and possibly amplification and scaling to produce an output of usable power and dynamic range, to drive such devices as audio speakers, lamps and stepping motors, which in turn generate the required environmental effects.

Although A/D and D/A converters vary greatly in their range of speed, precision, sensitivity, internal construction, cost, physical size, etc., there is little difference as far as functionality and system interfacing are concerned, and the software and hardware development for their utilization is fairly standard because of the simple and automated operation of these interfaces, which perform data conversion with little need for program intervention. We refer the reader to [1] for a more detailed introduction to analog/digital conversion, and will show in a later section a brief example of analog interface programming.

A possibility relevant to this section is that of multiplexing, in which a high-speed analog interface is shared by several external devices, each of which requires only a fraction of the available data rate. On the external side there will be a switching mechanism that connects each of the signals selectively to the analog interface, usually in sequence although other non-sequential selection processes are possible under either hardware or software control. Within the analog device a single digital data stream is handled, but this is segregated into individual streams in accordance with their data sources, either by the analog interface control program, or by the interface hardware which automatically deposits (for input) or picks up (for output) data to or from separate areas in memory.

(d) *Communication interfaces* A common type of I/O devices for microprocessor systems is the communication interface or line driver, for connecting these systems to data transmission lines over medium to long distances. Whereas most I/O interfaces are designed for short ranges, e.g. under 50 feet, often over multiple wire sets simultaneously carrying address, data and control pulse information, communication interfaces would transmit bits serially over a single cable in the form of groups of pulses. The serial bit streams will be separated into their original forms by the corresponding line receiver at the other end, which recovers the data despite occasional loss, distortion and interference over the link. Again there is a great variety of interfaces, having different speeds, distances and other performance attributes. The packaging or grouping of the data varies between different computers and between the interface types on the same computer, though generally they follow some agreed convention.

(e) *Other I/O devices* Because of the relatively low cost of microprocessor systems, they seldom have the major I/O devices commonly seen in large computer systems, such as card readers, lineprinters, disc or magnetic tape transports. More commonly a number of systems would share such a device, which may be linked to a special "service" system whose main function is to transfer data between the I/O device and the sharing systems. The service system, for example, might have a printer but not the others. When the latter have to produce a large program or data block stored on disc, one would simply remove the disc to the service system for output. Another possibility is that the service system would have several disc units so that disc-to-disc copies may be performed, while the remaining systems only have single-disc drives for memory-to-disc transfers.

Simpler hardcopy devices of lower cost and speed such as described earlier are now fairly common on microprocessor systems. Many systems also have simple and low-cost paper tape readers, since paper tape is still found as output medium for many measuring devices. Small magnetic tape cartridge units are also available for microprocessor systems, where they are used for backup storage and/or as a means of transferring data to and from other systems.

1.2.4 System Configuration

The configuration of processors, memory modules and I/O devices to form a complete system requires a considerable amount of development. The present subsection will be devoted to general comments on the overall structuring problem, and details of I/O interfacing will be deferred to Chapter 5.

We recall that a processor chip may only have a limited number of connection pins for the transmission of data, address and control/status signals. However, the operation of I/O devices requires a variety of control signalling information, some of which may be quite complex. Hence, the I/O control signals are usually not generated within the processor itself. Instead, the processor issues a limited number of signal pulses, merely to indicate what operation is needed for a particular device, and the I/O interface hardware accepts the "shorthand" command and produces from it the detailed control pulses required by the I/O device. Some microprocessor systems processors do not issue special I/O control pulses at all. Instead, the status control and data circuits in each I/O device are made to resemble memory locations so that we may perform the I/O operations by performing READ/WRITE operations on these pseudo-memory locations. Such systems are said to have *memory mapped* I/O. A printing operation might proceed as follows. The processor first performs a READ operation to fetch the status information contained in a register located within the printer, and tests the value to determine whether the device is free or busy. If the former, the processor would WRITE the binary number representing the symbol into the data register of the device. It would also WRITE a 1 into the START PRINT bit of the device control register, which causes the printer to relate the binary number to the specified symbol and print this. In this way, no special I/O control lines are required between the processor and the I/O device. Special circuits are, however, required in the I/O device interface to make the device simulate several memory locations, which makes the I/O programming more indirect and difficult to understand.

In a simple system the processor has complete control over the interface paths. It alone can place pulses on the address and the control lines to effect data transfers between itself and I/O devices or memory. Paradoxically, on more complex systems with microprocessors of greater power, control is diversified away from the processors. I/O may be able to pass data into and out of memory directly without involving the processor, and under some conditions they can even change the status of the processor itself, by placing information into some of its registers. These mechanisms need to be provided for by special hardware components in the interface subsystem and we will study them later. Figure 1.2 illustrates this, showing a slow I/O system which is fully controlled by the processor, but there is a separate fast I/O system which has its own independent path into the memory.

Despite the fact that most microprocessor systems need to have individually designed configurations and interface structures, a number of standardized interface systems are available to connect multiple devices of designated types and speeds to meet certain specifications concerning their data/control connections and signalling procedures. By designing devices to meet these specifications, manufacturers of equipment and processors provide simpler interfacing of their products into operational systems, and so improve the marketability of these products. In a later chapter we will be considering some of these standard systems, which include several designed to achieve computer resource sharing and data transmission.

1.3 THE STRUCTURE OF MICROPROCESSOR SYSTEMS: SOFTWARE

In considering the structure of microprocessor software we note that, except for very simple systems, the software used in a microprocessor system is likely to contain a large, even predominant, component of software provided by a manufacturer. We will be considering various aspects of this software in the following sections, commencing with the most important category of this software—namely that controlling microprocessor I/O operations.

1.3.1 Device Handlers

A device handler is a program segment which may be called upon to perform specified I/O operations on an I/O device. Because each type of device has its particular handling procedure, it is convenenient to prepare individual programs for each device handler and later integrate them into a complete I/O control system. Device handlers may be supplied by hardware manufacturers or prepared by the user in accordance with manufacturer specifications for the microprocessor.

Although device handlers may be written fairly easily, the integration into an I/O control system is a non-trivial task, because of the need for concurrent processing, arising from the mismatch in the speeds of the processor and the I/O device. With a terminal printer, for example, it takes the processor only a few microseconds to execute the necessary instructions to initiate an output, but a significant fraction of a second for the terminal to print the symbol, during which period the processor could have executed thousands of instructions. In consequence, the hardware of all but the simplest microprocessors provides for the overlapping of I/O operations with other processing tasks. As soon as a device handler initiates an I/O transfer on a device, it enters into a WAIT state to permit some other program, unrelated to the I/O device, to start execution. When the I/O device completes its operation, it may cause the other program to be suspended and the I/O device handler to be resumed from its WAIT state to initiate the next I/O transfer followed by a new entry into a WAIT state accompanied by the resumption of the unrelated program. Thus, we have two concurrent activities, each alternating between the execution state and the WAIT or suspended state. Since the second program may itself require to start another I/O operation on a different device it too may enter into a WAIT state and allow a third program, unrelated to either I/O device, to be started. In fact, for reasonable efficiency, it is desirable to establish a sufficient number of concurrent activities, to ensure that the speed of the processor will be fully exploited despite the need of the programs to wait for I/O completion.

A difficulty arises, however, where a number of programs have reached the stage where they are ready simultaneously to enter the execution state. We need to maintain information about all the concurrent activities and the status of each program and also to decide on a procedure which enables one program, out of those in the ready state, to be selected for execution. This is known as the *processor management function*. Additionally we need to be able to suspend a particular program to permit some more urgent task to be undertaken by the processor. This is known as the *interrupt function*. The two functions are related but not identical, the first being provided normally in software, and the second in hardware.

1.3.2 Processor Management

We have so far discussed processor management in connection with device handlers. In actual fact, any executable program, whether or not it is related to I/O operations, may be started as one out of several concurrent activities. Each such activity is called a *task*, and a system able to execute concurrent tasks is said to be *multi-tasking*. In large-scale time-shared computer systems there may be hundreds of concurrent tasks, the majority of which are executing interactively, in communication with user terminals, while other background programs are also being executed in the short idle intervals. In microprocessor systems the number of tasks is likely to be much smaller, but except for systems having very few I/O devices some form of multi-tasking is desirable to generate a mix of activities which will together cause reasonably high usage of all parts of the system.

Since the multiple tasks compete for execution by the processor, we need a *processor management routine* that selects one task for activation. At the time a microprocessor starts operation, the processor management routine is the first program to be started, and initially nothing else is active. Then external stimuli, such as users pushing buttons on terminals, will cause other programs to be executed. When such a stimulus is received, the processor management routine responds by locating the requested program in memory, or, if absent, by loading this in from disc storage, ensuring that the program is allocated the I/O devices and data it needs to do its work, and entering information about the program in a table (called a task control table). This table is located in memory and shows for each program its address, size, I/O device allocation, execution status, priority level, etc. The processor management routine then enters the inactive stage, allowing execution of the new program to start. However, as new external stimuli requesting other program executions occur, the currently executing program will be suspended and the processor management routine re-started, to bring in a new program, if the required resources in memory space, I/O devices, data files, etc., are available for allocation. Further external stimuli will generate additional tasks. Whenever the currently executing programme enters the WAIT stage or is unable to continue execution for any reason, the processor management routine is resumed to select, according to some priority procedure, another program listed in the task control table that is able to continue execution. The routine will also record the fact that one task has been suspended and the reason for the suspension, and the activiation of a different task, plus any other changes of the execution status of the tasks, in the task control table.

There is the possibility that an executing program may have a very long period of execution without any I/O operation initiations. It may be desirable to suspend such a program in order to allow other programs to be resumed or started which are more time critical. This capability is provided in the form of a unit called the *interval timer* which forms part of the processor. Before the processor management routine passes control to a program, it first loads a value defining the maximum permitted period of execution for the program into a register in the timer. During program execution this value is constantly reduced, and when it reaches zero, the processor automatically suspends the executing program to resume the processor management routine, which can then schedule some other program. A forced transfer of control away from the current executing program to some other program is called an *interrupt*.

Besides the timer, other devices such as I/O units and error detection components can also cause interrupts, as we will explain later.

1.3.3 Memory Management

We mentioned earlier that when a program is initiated sufficient memory space must be found for it. Similarly, when a program terminates the space it has been occupying should be freed and made available for use by other programs that will be starting execution later. A difficulty, however, lies in the size of the space required by a future program which is unlikely to be exactly the same as that available. If we have a block of unused memory sandwiched between two executing programs, then we could slot in a new program that is smaller than the free space, but certainly not one that is bigger. After slotting in a smaller program we would have some free space left over, which could be used to accommodate yet another still smaller program, or could be left unused until one program on either side of it finishes so that two free blocks could be merged into one to make it possible to accommodate larger programs.

Thus, we can identify two functions here. One is to record information about the parts of the memory. The current division into blocks, the free/occupied status of each block, its size and start/end points. The other function is to allocate and de-allocate memory blocks in response to requests, and to split and merge blocks as conditions warrant. These functions form the *memory management* for the system.

We have only considered memory management in the context of processor management and task start/completion. In actual fact, the scope of applicability is much wider because of the dynamic nature of the memory requirement of an executing program. In any average program there are parts which are executed for only brief periods of time during the life of the program, and of all the data handled by the program during its whole execution history only a portion would be actually needed at any particular moment. Thus, instead of having all the parts of a program and its data in the memory throughout its execution history, it would be more space-efficient to put each part of a program or its data into memory only when the part is actually needed. Earlier programmers achieved this through the use of *memory overlays*. Wherever possible the programmer divides his program and data into blocks which are never needed at the same time, and has several blocks share the same space in memory. A block which is no longer needed would be written out onto discs, and a new block needed later would be brought in to take its place. In this way, one can use only a small amount of memory to execute programs whose total space requirement is fairly large, and a skilful programmer can often take a sophisticated program written for a larger machine and overlay it sufficiently to make it execute on a microcomputer often with only a small overhead in disc READ/WRITE operations and space management manipulations.

In general, however, effective memory overlays are not easy to manage, and a more efficient technique provided by system manufacturers is to supply the software and hardware which incorporates the memory management function, making it unnecessary for users to concern themselves with the design of this function. Thus, when a program is initiated the management routine would allocate to it only sufficient space to contain those parts which are needed immediately. As execution proceeds and parts not currently in memory begin to be used, the system software would read the parts in from disc after allocating space for them. At the same time, it would be transferring out parts which are not currently being used when the space they occupy is required by some other executing program.

This scheme of dynamic memory management has some very fundamental implications, the main one being loss of control by the programmer over the exact location of data and program blocks. The memory management routine places these where it is most convenient. The programmer no longer knows what address a data item or a program instruction would have when the program is executing, because the actual address will depend on where the memory management routine puts the program or data block containing the particular item. This would in fact vary from run to run, or even vary within the same execution (since a temporarily unused program or data block may get dumped out, to be brought back into memory later but at some different address when it is used again.)

This is why in a processor implementing dynamic memory management it is necessary to distinguish between two kinds of memory addresses: those that appear in the machine program to identify *which* instruction or data item the program needs to access, and those, known to the memory management system but not to the program itself, which identify *where* in the memory the item actually resides. The former are called *virtual* addresses, and the latter *real* addresses or machine addresses. The "addresses" which appear in the program are not the numbers which will be sent to the address lines of the memory. Instead, there has to be some process of address translation to produce the machine address from the virtual address for every data item or instruction which is assessed during execution.

Let us consider virtual addressing in more detail, taking as an example the following simple system. The system permits concurrent execution of several programs, but each program, with all its data, is stored in one contiguous block, and there is no subdivision into smaller parts. (i.e. there is no dynamic memory management). From the program writer's point of view a given program may be assumed to be located in memory starting at address 0 and go up to maximum location M; i.e. the program's virtual space—the set of all its virtual addresses—is from 0 to M. But when a program actually executes the memory management routine may locate it anywhere in memory, say starting at address X. Thus, the actual space of the program is from X to $X + M$. If we wish to find the machine address of a data item whose virtual address is N, we simply add to N the actual starting address (or *base* address) X. So we have a simple address translation algorithm.

In the above case, "virtual" and "real" addresses simply differ by the addition of a base address value. But generally when one speaks of virtual versus real addresses one has in mind a memory management routine which subdivides a program and its data into smaller parts and holds in memory only those parts which are currently needed (which are said to constitute the *working set*). In such an environment a program is not stored in one contiguous block of memory, and virtual-to-real address translation is not simply the addition of the starting address. To identify any particular item, a virtual address has to identify in which program, in which part of the program, and in which item within the part it may be found. To translate the address it is required first to locate the position of the part, and then to add the initial address of the *part* to the *item* address within the part. While this might sound the same as the earlier case, it is actually much more difficult to implement. To start with, a program has several parts, and hence several initial addresses. Furthermore, the part that contains the required item may not be in the memory, in which case it is necessary to activate the memory management software to obtain space for the part, locate the part on disc, and load it into the allocated space. Only then could the address translation, the addition of the part's initial address to the item's address within the part, take place. So the address translation process may have three steps: a search to find if the data item is actually in the memory; if so finding the initial address of the block containing it; possibly a memory allocation/disc search/read process; and finally an address addition. Since this has to be done for every memory access, it is necessary to have special hardware which will perform the search and address translation very quickly. (The second part need not be particularly fast as it occurs only some of the time. Once a block has been read in items within it may be accessed repeatedly (Fig. 1.4).)

Systems that implement the above dynamic memory management process

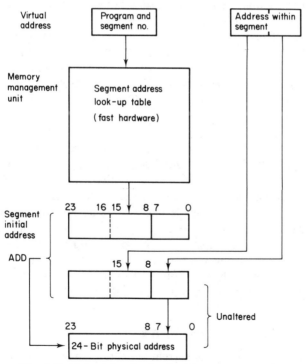

Fig. 1.4. Virtual to physical address translation.

are said to have *virtual memory*, but details can vary. In some systems programs are divided into blocks of a fixed size (e.g. 4096 bytes), regardless of whether the contents of each block naturally go together. These are called *paging systems*, and the fixed size blocks are called *pages*. In others programs are divided into their logical components which may vary in size, such as having each vector occupy a separate block. Such systems are said to have memory segmentation, and the logical components are called memory segments. Paging systems are easier to implement than segmentation systems, but are less efficient in memory utilization, since it may be necessary to bring in a whole page even when only part of it is required. Some large-scale systems implement paged segmentation, which combines two mechanisms to obtain the advantages of both, at the expense of more complex hardware.

A consequent benefit of memory management hardware is memory protection. With multiple concurrent programs it is important to ensure that each is only permitted to access those addresses that "belong" to it, and none is permitted to read or change data which belong to other programs. Such address checking can be built into the address translation mechanism. As programs only contain virtual addresses, which are in the form of program number/part number/item address, any explicit attempt by a program to access an item in a different program is immediately detected. However, this does not achieve complete protection, since a virtual address may have the correct program and part numbers, but the item address is too large, so that adding the initial address of the part produces a machine address which is outside the part and therefore belongs to someone else. However, this may be detected if the memory management hardware stores length information about each part and checks the item address against this before performing an address translation.

Few of the older microprocessors support dynamic memory management. However, both the Intel 8086 and Zilog Z8000 support memory segmentation, and the latter does so in a fairly sophisticated form using a special memory mapping unit to perform virtual-to-physical address translation. Some form of virtual memory is likely to be an integral part of the more sophisticated microprocessor systems currently under development to enable large-scale multi-tasking application programs and operating systems to be implemented, supporting multiple user and system virtual address spaces using only a moderately large physical memory (whose maximum size is limited by the number of address lines of the processor).

1.3.4 Device and Data Management

Another consequence of concurrent execution of multiple tasks is that a program may wish to perform I/O operations on a device currently being used by some other program. Similar problems may arise with multiple usage of the same data stored on disc, with the difference that is usually permissible for two programs to read the same data block concurrently as long as they do not change it, i.e. multiple read-only accesses are usually permissible, but not updates. Such conflicts can be solved in general only by the provision of yet more software components to control the allocation of devices and data blocks to programs.

A device management routine has a fairly simple task to perform. It maintains information about the devices under its control and requests for the use of each device. Requests to use a device are usually kept in the form of a chain with the earliest request at the head, or in a number of chains for different priority requests. A program requiring to use one of the devices would activate the routine, which may first check whether the program is authorized to use the device. If it is and the device is free, it is allocated to the program which proceeds to use it. If the device is already allocated to some other program then the new request is placed on the request chain and the requesting program put into the WAIT state. When the current user of the device finishes with it, the device management routine is again activated so that the device may be assigned to a waiting program or if none is waiting, left unallocated till the next request is received.

The main complication is the possibility of deadlock, which can arise whenever there are multiple users of multiple resources. Consider the situation where program A currently using device X, next decides that it also requires device Y. Device Y, however, is currently being used by program B and unavailable. So program A, having made the request, goes into WAIT state. At this time program B may decide that it too requires device X. Since this is currently allocated to program A and unavailable, program B will be suspended also after making its request. We now have two suspended programs, each waiting for the other to finish some work and to free its device, but since neither is able to continue work without the device it asked for, the processing cannot be performed and the devices may not be freed. Any further program which requests a device now allocated to either A or B would also be suspended because of device availability. We conclude that unless the software is equipped to detect the event and take action to resolve it (e.g. force program B to terminate and release device Y to A, and re-start B at a later time, we would have an intolerable situation. A solution is far from easy since we may have more complex forms of deadlock involving three or more programs, e.g. A has X and asks for Y, B has Y and wants Z, and C has Z but wants X, which may be extremely difficult to detect. We shall not discuss the issue further, but interested readers should refer to discussions on the deadlock problem in texts on operating systems or database management [3, 4].

1.3.5   Error Handling Routines and Utilities

Error handling routines are important to the user and should form part of the software available. I/O devices and memory modules can malfunction and produce wrong results, and some of the more common errors are detected by circuits built into the devices. This will lead to the activation of programs which attempt to recover from such errors, such as repeating an operation in the expectation that the error was only a transient one, or making an assumption of what the correct information should be. If recovery is not possible, the program that started the failed I/O terminal or memory operation may be terminated with appropriate error information appended to its output. There are also numerous possible software errors, such as unauthorized I/O or memory operations, attempting to read on a write-only device, asking for non-existent addresses, dividing a number by 0, arithmetic overflows (producing numbers which are too large to be storable in the memory). Again, appropriate routines may be activated to attempt to continue the program or to terminate the program originating the error. An activation of an error handling routine is a form of interrupt, since it is the forced transfer of processor control from the currently executing program which generated the error to another program, but such error interrupts are more often called *traps*.

There are also a number of frequently used service functions which are often provided by system manufacturers. Examples of these are file transfer routines for moving a block of data from one device to another; sort routines for rearranging data items and data groups (called records) in a large data set; loaders for reading in program and data and putting them into particular places in memory; dump routines for printing out contents of a block of memory for inspection; on-line debugging aids for observing the execution of a program under test.

1.3.6   Application Programs

With any microprocessor system the range of software to be developed by the system implementer and user varies, the main factor being the amount of system software support available from the manufacturer. Paradoxically, it is on the simpler microprocessors that the users have to do most because the software supply tends to be less. Not only are the system implementers responsible for the data analysis programs and I/O programs, they may even have to implement some form of resource allocation routine, to manage I/O devices, memory space and concurrent executions. Even for a strictly single-tasking system, there has to be some control program which will start and load successive programs and monitor and recover from error conditions.

However, with more powerful processors and better software support, many of the processor, memory and resource allocation routines are now standard supply, and so are device handlers for standard I/O facilities. Generally, users are only called upon occasionally to write their own device handler for special (especially user-developed) I/O devices, and their data analysis programs, usually called application programs because they relate to particular applications and also because they execute within the total structure of the system as "dependants" of the system software.

We shall make no general comments on the internal structure of application programs in view of their great variety. It is, however, useful to discuss how they fit into the overall software structure. Normally, data analysis programs are initiated by the processor management routine in response either to user requests entered from a terminal, or to the arrival of data at the data acquisition devices. There could also be several modules related to the applications which will be activated by different events. The following is a typical sequence of occurrences in a simple, highly manual system:

A. At system initiation a small control program is loaded and started. It is able to load other programs from disc in response to user commands. It resides in its private memory space.

B. The user requests that the data collection program be executed. This, including routines to handle the data acquisition devices and to build up files of data on disc, is loaded and started.

C. When the data collection completes, the data collector suspends itself and returns control to the control program.

D. The user may ask for a system program to be executed, which will accept commands from his terminal and return to him small portions of the data or status information about previous execution, so that he may see whether the data collection has been correctly performed. Loading such a program will, of course, re-use the memory space previously occupied by the data collector, which is erased and not available to be re-run without reloading the disc.

E. If everything appears to be correct, the user then requests the control program to load the data analysis program from disc. This erases his data inspection program.

F. After data analysis completes (presumably successfully, which the user may need to verify for himself, again using the data inspection program), the control program may then be asked to load a data dispatch program to send the data through a communication device to a remote machine, or to the printer for hardcopy output.

In such a system different program modules are simply stored individually on the disc; each is self-contained, having all the routines it needs already linked in. Such an arrangement wastes disc space, since it is necessary to put multiple copies of the same device handler into all programs that use it. However, program loads and erasures are simple.

On a sophisticated real-time system the various activities will be performed concurrently, under program control, rather than through constant human intervention. At system initiation a processor management routine will be started, and it will then initiate the data collector, analysis and dispatching programs to execute concurrently. At first, however, the analysis and dispatching programs will be in the WAIT state because the data they need are not yet ready, and the data collector is alone active. As soon as a data block is ready for analysis, the processor management routine will schedule the analysis program to execute in competition with the collection program, and after results of analysis begin to be available all three application programs will be active. While the collector is creating data block 3, the analyser will be working on data block 2 and producing a result block, while the previous result block, produced from data block 1 by the analyser, will be dispatched by the data dispatcher. Each of the three programs may call upon the same system routine to perform some standard system function, such as reading data from the disc, so that at any time one of them may be suspended to wait for I/O completion or device availability, allowing the other programs to control the processor. Obviously, the software structure for such a system is far more complex, but in return we have the capability of continuous and automatic real-time operation, with concurrent data acquisition, analysis and output and without human monitoring. Also, memory utilization is highly dynamic: for example, a device handler is loaded only when the application program asks for it.

1.4   Outline of Microprocessor Systems Development

In this section we shall give a brief sketch of the overall process of typical microprocessor systems development. Of course, details of each project differ, and system implementers have their individual preferences. However, some generally valid points may be identified and used as guides to system development.

Stage 1. System design   Before a project is initiated it is obviously necessary to identify clearly its aims and objectives. In microprocessor system development this takes the form of defining the characteristics and location of the input data and the form the output will take for a particular application.

The next step is functional design, in which the various functions the system needs to incorporate are identified as well as their inter-relations.

Easiest to identify are those functions directly related to data handling, such as device handlers, file transfer functions, and data analysis programs. Less easy to identify are the support functions, such as device/memory/processor management routines, since the need for them is clearly seen only after we have understood the likely structure of the data handling functions. A simple system with no concurrent processing, or a system in which various data handling functions are distributed among individual specialized processors, would have little need for processor management. A system with fixed allocation of memory space for programs (which may be all in ROM and therefore permanently fixed in memory) and data would require no memory management routines, and one in which I/O devices and data groups are permanently partitioned among different routines would need no device and data management. Usually, however, the deliberate restructuring of a system to eliminate the need for a system function often gives rise to the need for alternative support functions. For example, distributing device handling functions among separate processors immediately generates the need for inter-processor data transfer functions, whether via a common communication bus, or separate communication paths between each device-handling processor and the central data processing unit, or via the memory as a common storage element. In each, contention for the processor is replaced by contention for some other component of the system: in the first case different processors would compete for the use of the common communication bus to transfer data to some other processor.

Once the functions have been completely identified, some estimates of their requirements in terms of instruction execution, memory, and input/output traffic may be made, which in turn provice an estimate of the hardware requirements to handle the functions. Possible alternatives for carrying these out either by software, use of a separate smaller microprocessor, or by special-purpose hardware, may then be considered. Decisions made in this step may well affect the estimates made in the earlier stage, and some iterative refinement may need to be performed. Eventually a system configuration is determined that provides for all the functions previously identified, whether in the form of hardware or of software. This completes the design process. However, during the implementation stage errors in design estimation may well come to light, in which case modifications in the configuration may need to be incorporated and the procedure outlined above reiterated to arrive at a new system structure design.

Stage 2. Hardware development   If the hardware configuration determined in the design stage conforms to a standard structure supported by the microprocessor manufacturer, then little hardware development is needed. Even where the manufacturer does not deliver complete assembled systems, it is likely that the development work may be confined to interconnection of fabricated components and interface modules, supplemented perhaps with a few locally designed circuit boards.

More substantial work is needed in those cases where it is necessary to incorporate non-standard components and I/O devices into the system, particularly user-developed modules for special purposes, such as modules for high-speed vector processing, image processing or bus extenders for connecting additional I/O devices or memory modules. Many suppliers of such devices may offer interface modules which will make the device appear to be a standard one to a given system. That is, the interface modules will receive and generate control, address and data pulses on the system interface bus in accordance with the standard requirements of the system, to enable appropriate data transfers to occur between the microprocessor system and the device. But often the development of such interface modes may need to be carried out by the user.

This illustrates that the process of hardware design is primarily one of identifying a set of distinct functions and their inter-relations, with the aim of specifying the complete system in terms of a number of functional modules, each to be activated under certain conditions (defined in terms of received control pulses), and whose activation will cause a prescribed set of input pulses to be received and certain output pulses generated from them. Once the functional modules and the input/output relation of each of them have been clearly defined, it is possible to proceed with the specification of the internal circuits of each module using well known techniques of Boolean logic design. While the overall process may be complex and time-consuming, the principles involved, which will be discussed in the following two chapters, are fairly simple.

The difficulties of interface design are considerably reduced by the adoption of a standard hardware configuration, or a configuration based upon one of the standard interfaces, such as the RS232 interface. These have been designed after careful consideration of the requirements of different applications, with the circuits incorporating a number of common control and status lines, and the user given a choice of several control sequencing procedures that satisfy the needs of a large range of devices. A user designing his special purpose equipment to follow a standard interfacing procedure would minimize interfacing incompatibility for very many units. We shall be considering standard interfaces later.

Stage 3. Software development The software development stage proceeds concurrently with hardware development. It is not necessary, nor is it desirable, to wait for the completion of the latter before starting the former. This would take too long, and problems encountered during the software development stage may well call for substantial hardware modifications. Once the hardware has been designed *functionally* it is possible immediately to start writing programs for the devices, at least at a conceptual level.

At this level, development of both hardware and software requires the identification of individual system components having distinct functions, their inter-relations, and the input/output variables of each component as well as the condition for its activation. In a complex and large system, each component may be further subdivided into smaller parts in a similar way, until we arrive at a set of components which are simple and sufficiently well understood that they may be specified individually in terms of either logic components (hardware) or programming instructions (software). However, while the similarity in approach may be quite pronounced at an early stage of development, in later stages the two development processes may be very different, reflecting not only the difference in the basic building blocks and the tools employed but also in the type of structures that are implemented.

First, the operation of a software system is fundamentally sequential. Execution starts in one particular block, which processes some data and uses the result to determine which software module is next to be executed. Similarly, the second module would either pass control to a third module or perhaps return control of the processor to the first module, which will then select some other module for execution depending on the results of previous executions. This is the type of behaviour one experiences, for example, in processor management, and any system with such behaviour would be more suitable for software rather then hardware implementation. While concurrent processing of multiple software modules is common, this is due largely to the need to incorporate device handling, that is, the handling of *hardware*, and even here the individual device handling involved is sequential. Certain types of data analysis may involve vector parallel operations and the sequential type of data manipulation we have been describing will not be suitable for this. Special purpose hardware implementation may be preferable and an example of such a device will be discussed in Chapter 3.

A notable difference between hardware and software development is the relatively greater freedom and flexibility of the latter. In hardware construction there is usually a final design which is clearly the most economical and effective for meeting the functional requirements, whereas a program for a given job can usually be written in several alternative ways with little difference in the resulting performance.

A further consequence of the greater flexibility offered by software systems is that tasks which are subject to later modifications are generally implemented in software. Inserting one or two extra instructions into a program and re-translating it into machine code is a fairly trivial task, whereas adding a new component or connection to an existing circuit board may be extremely difficult because of space and other limitations.

In a large-scale software development project, high-level languages should be employed wherever possible since this considerably improves programmer productivity. This implies the availability of a suitable compiler which will translate the high-level program into the machine language of the target microprocessor. However, certain types of operations such as repeated manipulation of a small set of data items stored in registers, are much slower if implemented in high-level languages. For similar reasons I/O operations and device handling are difficult to program in such languages as these tasks generally depend on testing and manipulating bits in device registers. Often, therefore, the system design is arranged to separate out those parts which are particularly time critical, to be executed frequently, for example, or those directly engaged in register manipulations, and to program them separately in microprocessor assembler or machine code.

Stage 4. Testing The testing stage of a microprocessor systems project generally consists of three phases. The first phase, module testing, occurs during the latter part of hardware and software development, when individual hardware and software modules are tested to determine if their input/output behaviour conforms to design specification. Hardware modules are tested using pulse generators to provide a simulated input and their output displayed on oscilloscope screens and LED display panels. (We shall consider this further in Chapter 6 when we discuss development systems.) Software modules are harder to test, since they generally have input/output in the form of data passed from and to other software modules. A special test driving program is used which will prepare suitable data blocks in memory or on disc and activate a module under test such that it will fetch the prepared data blocks as input, and produce output for retrieval by the driving program, which then prints the output for inspection. The amount of work expended on producing a working test driver is frequently a significant portion of the programming effort.

As development and module testing proceed side by side, the stage is reached when some of the modules may be linked together into a larger part, which may again be tested on its own to verify its input/output behaviour. When all the modules have been verified, the linking into a complete system may take place. Phase two of the testing stage then occurs, with the complete system testing in an "artificial" and controllable environment, usually called a work station, which we describe later in detail. This consists of a microprocessor of the same model as will eventually execute the software in production, but having additional memory and I/O devices to facilitate program modification and testing, such as discs for program storage and fast loading, interactive terminals for on-line debugging and a printer for dumping memory contents for verification. Alternatively, it may have a more powerful processor which can simulate the instruction set of the target microprocessor, with the extra processing power devoted to test programs and device control functions running concurrently with the program being tested. The final phase of testing places the system with complete and wholly integrated hardware and software components in its working environment connected to its controlled device (machine or physical system). The total system behaviour is then observed in response to typical environment conditions. This phase may extend over a considerable period after the system has gone into production, to increase confidence and obtain estimates of system reliability.

Stage 5. Maintenance and documentation Because a working microprocessor system, especially its software, is a fairly complex object, and often the product of team work with each person responsible for only a small portion, it is not unusual that no one person wholly understands the system, and that the system contains errors which do not affect observable behaviour until a special combination of circumstances occurs. It is therefore important that the structure of the system be described clearly in detail in accompanying documentation so that should any part need to be modified, whether to eliminate error or for some other purpose, it would be possible for a programmer previously unfamiliar with the system, or at least unfamiliar with the part being modified, to understand it and make the correct changes. In the design phase a number of such documents are produced to define the system structure in general terms, and individual programmers insert comments into programs to aid comprehension and memory. In a well organized project, little more is needed to produce useful documentation other than a systematic re-organization of such material at the end of the project to incorporate subsequent changes and to make the material comprehensible to persons not involved in the project itself. Other required documents are user manuals to explain system operation from an outsider's point of view, without involving internal system structure, and system updates for explaining changes and improvements or to point out errors discovered by users of the system.

References

1. Beauchamp, K. G. and Yuen, C. K. (1980). "Data Acquisition for Signal Analysis". Allen & Unwin, London.
2. Millman, J. (1979). "Microelectrics: Digital and Analog Circuits and Systems". McGraw-Hill, New York.
3. Deen, S. M. (1978). "Fundamentals of Data Base Systems". Macmillan, London.

4. Lister, A. M. (1979). "Fundamentals of Operating Systems" 2nd Edn. Macmillan, London.
5. Laventhal, L. A. (1978). "Introduction to Microprocessors: Software, Hardware Programming". Prentice-Hall, Englewood Cliffs, N.J.
6. Fabian, M. and McLaughlin, R. (1981). "An Introduction to Mini and Micro Computers". Peregrinus, Stevenage, U.K.
7. Osborne, A. (1981). "Sixteen-bit Microprocessor Handbook". McGraw-Hill, New York.
8. Osborne, A. (1981). "Four and Eight-bit Microprocessor Handbook". McGraw-Hill, New York.
9. The 8080/8085 Microprocessor Book (1980). Wiley, New York.
10. McGlynn, D. R. (1980). "Modern Microprocessor Design: Sixteen-bit and Bit-slice Architecture". Wiley, New York.

Digital Electronic Devices

2.1 LOGIC GATES

In the previous chapter we mentioned that microprocessor hardware is constructed from switching circuits. In this chapter we shall present the basic principles of switching circuits, and see how some simple computer components may be constructed from them. The design of more complex devices will be the subject of Chapter 3.

The basic electronic switch, shown diagramatically in Fig. 2.1 connects or disconnects the two conductors 1 and 2 in accordance with the logical state of the signal applied to the controlling line a. If this signal is at a high voltage level (logical 1) then the switch is closed or turned on and conductors 1 and 2 are effectively in contact. If it is in the low voltage level (logical 0) then the switch is open or turned off. To use the switch we apply a voltage level V to one of the conductors, as shown in Fig. 2.2 and study its input/output relation.

We see that, if the input a is high (1), then the switch in the figure would be turned on, and the output x will receive a high voltage (1). If the input a is low (0), no electric current flows through the switch to the load so that the output x, is therefore at the 0 level. In short, the input/output relation may be described quite simply as $x = a$. A slightly different device is shown in Fig. 2.3. Here $a = 1$ would turn on the switch, placing output x at the 0 level, whereas $a = 0$ would connect x to the high voltage or 1 level. We see that the output x is the *complement* of input a, being 0 when the input is 1 and

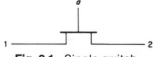

Fig. 2.1. Single switch.

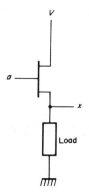

Fig. 2.2. Simple device.

vice versa. The device of Fig. 2.3 is called an *inverter*, and its input/output relation is notationally represented as $x = \bar{a}$, where ‾ indicates the process of inversion.

We now examine the device shown in Fig. 2.4. It may be seen that if $a = 1$ a current flows through the load and the output is at the 1 level. Similarly, if $b = 1$, x would be 1. And if both a and b are 1, x would again be 1, with half the current flowing through each switch to the common load. (The total is limited by factors other than the switches, and the fact that there are two connected paths does not necessarily mean that there would be a higher voltage output.) However, if both a and b are 0, then x would be 0 since no current passes through the load at all. To sum up, the output is 1 if either *a or b* is 1. Because of this, the device shown is called an *OR gate*. Generalizing to

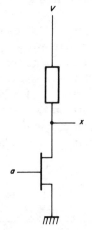

Fig. 2.3. Inverter.

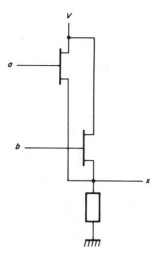

Fig. 2.4. OR gate.

gates of more than 2 inputs, an OR gate outputs 1 if *any* of its inputs is (are) 1, but 0 if *all* its inputs are 0.

If we now consider Fig. 2.5, we see that output *x* is 1 if both *a and b* are 1, and is 0 if either or both are 0. This device is called an *AND gate*, Together, the AND gate, the OR gate and the inverter form a complete logic system. That is, by connecting these gates appropriately one can create any device with a given binary input/output relation.

Before we discuss the issue of device design, it is first necessary to introduce a number of notational systems. First, how do we specify the input/output relation to a device we want to design? A commonly used method is the *truth table*, which exhaustively lists the required output values for every possible combination of input values. Take, for example, the addition of two one-bit numbers. The device would have two inputs, *a* and *b*. If $a = b = 0$, then the output is 0. If $a,b = 0,1$ or $1,0$, the output is 1, and if $a = b = 1$ the output is 2, or 10 in binary notation. We see that the device needs two output lines, say *x* and *y*, and the input/output relation may be exhaustively listed as follows.

Table 2.1. Truth table for addition of two one bit numbers.

a	b	x	y
0	0	0	0
0	1	0	1
1	0	0	1
1	1	1	0

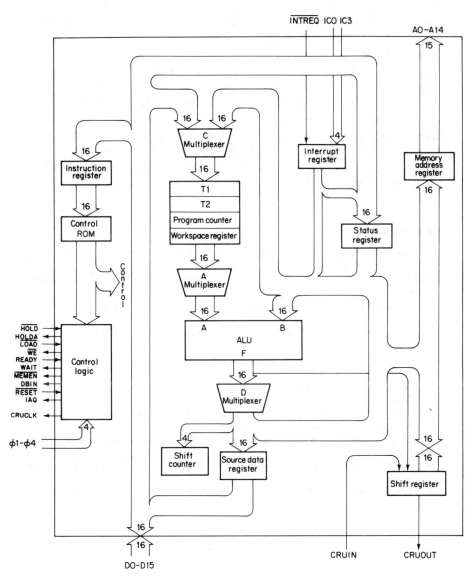

Fig. 3.22. Texas Instruments TMS9900 processor. (With acknowledgments to Texas Instruments Ltd.)

buses that pass information between the processor and the rest of the micro-processor system (The A and D paths in the figures. Note that the data bus is 8 bits wide for Intel 8085, but 16 bits for the Texas 9900). Timing and control logic can be seen in both designs, where its task is to generate status and control information pulses for transmission to external units and to accept external requests (e.g. READY, HOLD) from these outside units. Various registers, the arithmetic-logical unit, and instruction decode/execution control units may also be identified. A large number of internal data paths are present, which are controlled by the execution control unit through various lines not illustrated. While at first sight the figures might seem extremely complicated, but once we have identified the main components they become more informative, and it begins to emerge that the general structures of the two diagrams are quite similar.

References

1. Mano, M. M. (1979). "Digital Logic and Computer Design". Prentice-Hall, Englewood Cliffs, N.J.
2. Kline, R. M. (1978). "Digital Computer Design". Prentice-Hall, Englewood Cliffs, N.J.
3. Booth, T. L. (1978). "Digital Networks and Computer Systems" Wiley, New York.
4. Bartee, T. C. (1981). "Digital Computer Fundamentals", 5th Edn. McGraw-Hill, New York.
5. Wallace, C. S. (1964). A suggestion for a fast multiplier, *IEEE Transactions on Electronic Computers* **13**, 14–17.
6. Mead, C. and Conway, L. (1980). "Introduction to VLSI Systems". Addison-Wesley, Reading, Mass.
7. Yuen, C. K. (1980). A bit-serial device for extracting a vector element of a specified rank, *Digital Processes* **6**, 207–210.
8. Yuen, C. K. (1980). A bit-serial device for maximisation and sorting, *Proc. IEEE* **68** 196–197.
9. *The TTL Data Book for Design Engineers* (1980). Texas Instruments, Dallas, Texas.
10. *9900 Family Systems Design and Data Book* (1980). Texas Instruments, Dallas, Texas.

Microprocessor Programming

4.1 OVERVIEW OF MICROPROCESSORS

In an overview of microprocessors one inevitably encounters the problem of defining what is or is not a microprocessor, and which of the numerous semiconductor components with processing capabilities appearing in the recent past may be considered the earliest microprocessor. The early "programmable logic elements" or "intelligent controllers", having a small memory for a user-defined program, may well be considered as simple microprocessors.

However, because of the restricted capabilities of such devices, their impact on the general computing scene was small. It was only when processing elements capable of addressing a fairly large program and data memory came into being that larger more complex programs processing reasonably large amounts of data became possible and more elaborate device interfacing requirements could be accommodated. From this point the microprocessor began to assimilate many of the tasks previously carried out by the mini-computer and an increasingly wide range of controlling tasks not previously considered economic in computer terms.

The early microprocessors operated with very short wordlengths, one example being the Intel 4004 which was a 4-bit machine. Some of these units contained *bit slicing* capabilities permitting the construction of processors of larger wordlengths by appropriate interconnection of short-wordlength devices. Alternatively, data of greater wordlengths, such as text (consisting of 8-bit characters), or numerical values of low precision (e.g. 16-bit integers) could be processed by writing the necessary software to simulate longer wordlength processing. However, because of the required cost of extra hardware or software, it was feasible to apply such devices only to fairly special applications. In this sense the invention of the first 8-bit microprocessor, the Intel 8008, was a major step; although quickly superseded by the more powerful 8080 series of processors which we consider later. The 8008 was for its time an extremely versatile device, with an extensive instruction set previously found only with mini-computers including the ability to process 8-bit data without software simulation together with a reasonable set of facilities to implement such program-constructs as loops, conditional jumps and subroutines. Even today some of its design concepts remain interesting from an educational point of view.

Table 4.1 provides an extensive, but far from complete, list of 8-bit and 16-bit microprocessors.

Although many manufacturers are or have been concerned with the production of microprocessors, three have made major contributions to their development. These are Motorola, Intel and Zilog. Intel was one of the earliest of the pioneer developers and has since produced a large range of microprocessors with the 8-bit 8080 and 8085 applied quite widely and the later 16-bit 8086 rapidly gaining in usage and importance. The Motorola 6800 processor, also an 8-bit device of wide application, provides an interesting contrast to the Intel 8080, being based on very different design concepts, and these were further developed to produce the 16-bit Motorola 68000 model. In contrast, the Zilog Z-80 was designed to be upward-compatible with the Intel 8080, such that all the instructions of the latter are also available on the former. The Zilog Z8000 is a fairly sophisticated 16-bit microprocessor, and is free from many of the limitations of earlier products, though its architectural concepts still show some relation with the earlier Intel and Zilog products.

The wide range of microprocessors makes it impossible for us to study each one individually. Thus, we shall base our discussion in the Intel 8080/85, but paying constant attention to the basic programming ideas in our discussion such that the reader may readily apply the concepts to other microprocessors of different design. However, in view of the wide use of the Motorola 6800 and the importance of its particular architectural ideas, we shall also be considering these ideas in comparison with the 8000 series developments.

A major difference between Intel and Motorola design architecture lies in the use of its registers. We saw earlier that each processor contains a number of registers for storing values which are frequently needed during program execution. Some of these values may be *data*, but there may also be *addresses* which refer to particularly important memory locations. Thus, at an early stage of processor design one is faced with a fundamental question: Should there be two different types of registers, one for storing addresses and one for storing data, or should there be *general purpose registers* which may be used to contain both types of information? Each approach has its particular advantages. In 8-bit microprocessors the basic data unit is a byte, which is too short for addresses since a byte can only refer to 256 (2^8) different locations and addresses are usually 16 bits in length. It would thus seem sensible to have

Table 4.1. List of 8- and 16-bit microprocessors

Manufacturer	Device	Technology
8-Bit microprocessors		
Fairchild	3850	NMOS
	3859	NMOS
	3860	NMOS
Intel	8008	PMOS
	8080	NMOS
	8035/8039	NMOS
	8041/8741	NMOS
	8048/8748	NMOS
	8085	NMOS
	8088	NMOS
MOS Technology	6502–6	NMOS
	6512–15	NMOS
	6800	NMOS
Mostek	3870	NMOS
	3871	NMOS
	3874	NMOS
Motorola	6800	NMOS
	6801	NMOS
	6802/6808/6846	NMOS
	6803	NMOS
	6809	NMOS
National Semiconductor	ISP8A600(SC/MP-II)	NMOS
	INS8040	NMOS
	INS8060	NMOS
	INS8070	NMOS
	16008	NMOS
RCA	1802	CMOS
	1803	CMOS
Rockwell	PPS-8	PMOS
	PPS-8/2	PMOS
Western Digital	1621	NMOS
Zilog	Z80	NMOS
	Z8	NMOS
Advanced Micro Devices	29116	Bipolar
Data General	mN601	NMOS
	mN602	NMOS
Fairchild	9440	Bipolar
	9445	Bipolar
Ferranti	F100L	CDI Bipolar
Intel	8086	NMOS
Motorola	68000	NMOS
National Semiconductor	PACE	PMOS
	INS8900	NMOS
	NS160016	NMOS
	NS160032	NMOS

(cont.)

Table 4.1. *(cont.)* List of 8- and 16-bit microprocessors

Manufacturer	Device	Technology
Signetics		Schottky Bipolar
Texas Instruments	TMS9900/9940	NMOS
	TMS9980/9981/	NMOS
	TMS9995	NMOS
	SBP9900	I^2L
Western Digital	WD-16	NMOS
	Pascal Microengine	NMOS
Zilog	Z8000	NMOS

Bit-slice microprocessors

Manufacturer	Device	Technology
Advanced Micro Devices	2901	Schottky Bipolar
	2903	Schottky Bipolar
Fairchild	9405	Schottky Bipolar
	34705	CMOS
	100220	ECL Bipolar
Intel	3002	Schottky Bipolar
Motorola	10800	ECL Bipolar
Texas Instruments	SBP0400	I^2L
	SN54S/74S-481	Schottky Bipolar

longer registers for addresses and shorter ones for data. Further, the ways in which we use the two types of information are different. For example, since the contents of address registers are used in memory READ and WRITE instructions the registers are frequently connected to the address lines of the microprocessor, whereas the data registers are more likely to be connected to the data lines. The types of manipulations one wishes carry out on addresses are also different from those performed on data; for example, one hardly ever wants to multiply two addresses. Hence, in theory at least, having two types of registers should be more efficient, permitting the designer to build only address-related capabilities into the address registers and data-related facilities into data registers. Also, the separation should produce a more efficient instruction format. With general purpose registers an instruction has to specify *which* register to use plus *how* to use it, i.e. whether as an address or data register. In contrast, with specialized registers much of this kind of information is implicit, since the contents of address registers can only be used as addresses and those of data registers only as data, and many address-type operations can only refer to address registers and other data-type instructions refer only to data registers. Further, with specialized registers, a processor would contain only a small number of each type: e.g. the Motorola 6800 has two 16-bit address registers and two 8-bit data registers, and it takes only one bit to specify which address register one wishes to use and another bit to specify which data register. In contrast, the Intel 8080 has seven 8-bit general purpose registers, and it takes 3 bits to specify which register one wishes to use and some additional bits to specify how to utilize the value in the register (we shall explain how in more detail later). However, experience has shown that specialized registers do *not* necessarily lead to greater processor efficiency. Other factors come into play. First, with general registers the programmer can devote more of them to data or address as needs arise. Second, addresses often have to be computed through arithmetic or logical manipulations. With specialized registers the programmer is forced to load address values into data registers, manipulate them there, and finally load the results into address registers for use in memory accesses. Thus, the separation of register types tends to produce certain processing inefficiencies. This is why, while each type of processor has its adherents, both have been able to find their place in a large number of applications, and neither may be said to have shown itself superior to the other. A second point of difference between Motorola 6800 and Intel 8080 lies in the handling of I/O operations. The former uses memory-mapped I/O (see Chapter 1), while the latter has special I/O instructions, which we shall discuss in a later section.

It is also worthwhile to explore briefly the more modern processor models and point out their salient features. The Z8000 has 16 general purpose registers of 16-bit wordlength. Each register may be used to hold data or address information in several different ways (called *addressing modes*, to be discussed later in other contexts). Provision is made in the processor design for memory segmentation through an attached memory mapping unit used to convert virtual addresses passed from the processor into physical addresses before placing them on to the address bus of the I/O and memory system. Up to six virtual address spaces, each of which may be subdivided into 128 segments of maximum size 2^{16} bytes, and a maximum of 2^{24} bytes of physical memory, are permitted by the processor design. This allows a high degree of dynamic multiprogramming to be carried out with different programs contained in separate virtual spaces and with the program and data blocks in each being dynamically loaded into the physical memory. In many ways the Z8000 design compares favourably with the best minicomputers and mainframes currently available, the main limit to its wide utilization in application being the less comprehensible availability of software and poorer I/O system capability.

The Motorola 68000 provides an interesting contrast to the Zilog design. Like the earlier 6800, it has separate data and address registers, but with much increased power, having 8 of each type both 32-bits in length. The instruction set provides a high degree of flexibility in register utilization through various addressing modes and manipulations of 1,8, 16 or 32 bit data types are available in various forms. Potentially, the machine architecture is capable of supporting fairly large systems and complex processing tasks. However, current hardware products implement only a subset of the design. For example, the address bus is currently limited to 24 bits rather than 32, though there is the possibility of a 3-bit extension through hardware segmentation, i.e. having several physically separate address spaces distinguished from each other by additional address lines. It is expected that memory mapping hardware and vector manipulation instructions, analogous to those already available on Z8000, will be implemented, which will greatly improve the flexibility of the architecture.

The Intel 8086 is an earlier 16-bit design than the two we have just considered. Its architecture is limited by the small chip size (allowing only 40 pins in comparison to 48 and 64 for the Z8000 and 68000 respectively) and by the designer's decision to maintain a degree of software compatibility with the 8-bit 8080 design. The seven 8-bit registers of the 8080 are retained with an 8-bit extension added to the main data register (accumulator). Two sets of address registers, four of these for normal data access called stack, base, source and destination respectively, and another four for memory segmentation purposes, are added, and the instruction set and available addressing modes are much expanded over those of 8080. The data bus width is doubled to 16 bits, and manipulation of multi-byte data is provided for in the instruction set. However, as only 20 address pins are available, the processor is less able to support complex configurations. Nevertheless, the processor and its smaller derivative the 8088 have proved to be a useful addition to the Intel range, mainly because of improved software availability and the fact that much 8080 software can be modified without difficulty for the new processors.

Of the other manufacturers, the MOS Technology 6502 processor is used in many small personal microcomputer systems, and its architecture bears some similarity to the Motorola 6800. The various models of processors from the instrument manufacturers are also used widely, especially in other products of the company supplying the processor concerned. A number of high-speed bit-slice chips, based on TTL or ECL technology rather than MOS, are also available. As such components have much lower chip density, they are particularly suitable for applications in which high processor speed is essential but elaborate instruction sets are not needed. Some examples of special purpose bit-slice components will be discussed in Chapter 7.

Many 32-bit processors are currently in development stage. A particularly important feature of the current situation is the integration of hardware development with the new methodology of software engineering, in particular the new, multipurpose language ADA. However, discussions of these developments are beyond the scope of this present volume.

4.2 INTRODUCING THE INTEL 8080 MICROPROCESSOR

We are now ready to start a more detailed study of the Intel 8080 and 8085 processors and their programming. Since the 8080 exhibits most of the essentials of microprocessor design and operation it is useful to consider it in detail here, although we will find it necessary from time to time to compare its facilities with the later 16-bit microprocessor, the Intel 8086.

Fig. 4.1(*a*) shows a sketch of its hardware configuration. The processor is independently packaged in a 40-pin chip, Fig. 4.1(*b*). Sixteen of these pins are address connections, and will be joined to the address lines of any microprocessor system containing the device. Eight pins are data connections, and will be joined to the data lines of a system. The remaining sixteen pins are for such essential connections as power supply, earth, start pulse, interrupt signals, memory WRITE control, etc. Because of the shortage of pins, certain required control pulses do not have their own connectors. For example, the chip does not provide a separate pin for the memory READ control pulse, which when required, has to be generated by adding extra circuits. (We shall consider such circuits and similar problems in the next chapter.) With 16 address lines, an Intel 8080 processor is able directly to address up to 2^{16} memory locations, each location containing a byte. These are provided for by interfacing one or more memory chips to the processor by appropriate connection of address, data and control lines.

We may note from Fig. 4.1(*a*) that the processor contains a number of components carrying out various functions, such as an arithmetic/logical unit which is used to perform data manipulations, an instruction decoder and execution control unit which analyses instructions fetched from the program memory and generates the necessary control pulses to effect the execution of these instructions, together with a number of registers. Of these, the seven registers marked A, B, C, D, E, H and L are available for storing data and addresses, whether constant or variable. Instructions are provided in the design of the processor to move data from one register to another, from registers to memory or vice versa, or to combine the contents of two registers arithmetically or logically (but with these instructions generally one of the two registers has to be register A. One cannot, for example, add the contents of registers B and C. Because of the special status of register A it has the special name of *accumulator*, since the results of a series of arithmetic operations are accumulated in register A.) We shall study these instructions and show their application to the solution of simple problems later in this chapter.

Figure 4.1(*a*) also shows a number of additional processor registers that service other facilities. The *instruction register* (IR) is used to contain the instruction which is currently being executed, and we may note that it is connected to the input path of the instruction decoder, so that the latter may

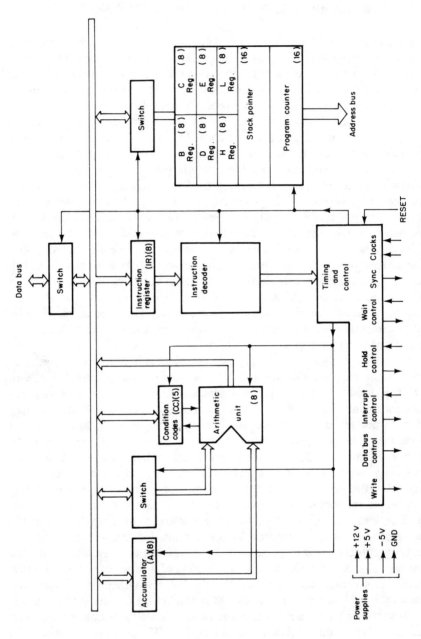

Fig. 4.1. (a) Intel 8080 architecture (with acknowledgments to Intel Corp.).

(*b*) Intel 8080 chip pin configuration.

analyse the instruction to determine what processing is required. By storing the instruction in a register, the design ensures that the information is held throughout the whole duration of the instruction execution and is constantly available. For similar reasons, an address buffer register and a data buffer register are provided to ensure adequate control of the interface circuits. There are also some temporary registers, which are used to store intermediary results produced during instruction execution and needed later to compute the final result. These registers are not available to the programmer for the storage of information and are said to be *transparent* to him. The remaining three registers, the *program counter* (PC) the *stack pointer* (SP), and the *condition codes register* (CC) play a very direct and essential part in the programming of microprocessors, and it is vital that a programmer should understand their functions and how to make use of them effectively.

4.2.1 Program Counter

The primary function of the PC is to store the address of the instruction *about* to be executed. When we wish to execute a program from the program memory, the starting address of the program is loaded into the PC and a START pulse sent to the processor. This causes the processor to initiate an instruction fetch, which it performs by connecting the content of the PC to the address lines and initiating a memory Read signal. Because the content of the PC is the address of the first instruction, the memory READ would then bring the instruction from memory. By connecting the data lines to the instruction register, the processor is enabled to send the instruction read out from memory to the correct place for decoding and execution. After the instruction has been fetched, the processor increases the value held in the PC, so that it contains the address of the second instruction of the program. When the execution of the first instruction is completed the processor is then able to use the content of the PC to fetch the second instruction, again increasing the value held in the PC to refer to the third instruction, and so on. We see that, by following this regular procedure the execution control will proceed from one instruction to the next through the program, at all times with the PC pointing to the location *below* the instruction currently being executed, i.e. to the next instruction, *if* the program contains nothing but a continuous sequence of instructions.

We have emphasized the word *if* deliberately, for it is possible to place some data of address information within the instruction sequence and indeed this mixing of several types of information has certain important uses. For example, some of the instructions in the program may need one or more constants, which would have to be fetched from some place into which they have been previously stored. Now one could of course reserve a number of registers for this purpose. However, the number of constants one can accommodate in this way would be rather limited, and it would be extremely inconvenient continually to maintain information of which register contains which number. Instead we store each number with the instruction that will be using it, say immediately below the instruction in the program memory. But how do we fetch the number when the instruction is actually executed? This turns out to be very easy. We know that the PC always points below the instruction currently being executed, so that, if the processor places the content of the PC on the address lines and performs a memory READ operation, it will fetch the constant required by the instruction. After this has been done the processor would again increase the value of the PC to ensure that it refers to a memory location further down, which contains the next instruction. The process is illustrated in Fig. 4.2.

To summarize:

(a) The programmer has the option of putting some of the data required during program execution immediately below the instructions that use the data.

(b) During execution the PC points successively to each location in the program memory, and its content may be used to fetch each instruction as well as the data used by that instruction.

(c) Every time we use the PC its value is generally increased by 1 to make it refer to the memory location immediately below the one previously accessed. In this way, one ensures that the PC will point to the data used by an instruction after the instruction has been fetched, and that

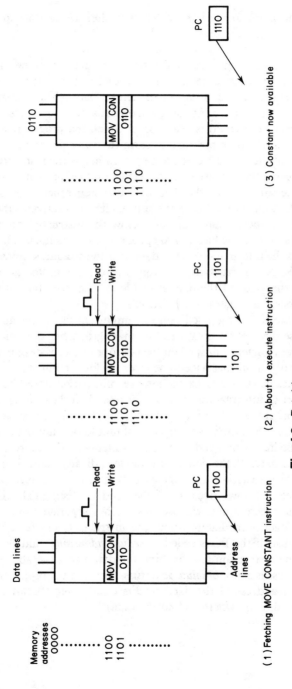

Fig. 4.2. Fetching constant data using the PC.

it will point to the next instruction after all the data for the current instruction have been fetched.

Unfortunately, only particular types of "data" may be mixed within the program; the programmer has to know their values at the time the program is being written, e.g. constant coefficients. Data whose values are computed during the execution of the program obviously cannot be handled in this way. An alternative method is available in which the programmer reserves some locations in the data memory to store the values of such variables, and the *addresses* of these data locations are placed below the instructions that use the data. During the execution of an instruction having an attached data address, the processor accesses the data in a two-step process. First the processor places the value of the PC on the address line and activates the READ signal. This causes the memory location below the instruction to be read and the address stored in the location to appear on the data lines. The processor then transfers the information on the data lines to the address lines, and activates a second READ instruction. The contents of the data storage location are read making the value of the data available for the execution of the instruction. This process is illustrated in Fig. 4.3.

We have actually glossed over one important fact in the above description. Addresses are nearly always 16 bits in length, while data may be either 8 or 16, or even longer. Thus, when the microprocessor is designed, some parts of the instruction format have to be reserved for the purpose of *addressing mode* specification. Does this instruction use information stored below it? and if so, what sort of information? Is it an address? Is it data? If so, how long is the data item? If the addressing mode indicates that there is an address below the instruction, then the execution control unit knows it must read the two bytes below the instruction and place their contents on the address lines in order to fetch the data. If the addressing mode indicates there is one byte of data below the instruction, then the execution unit would read only the location immediately below. Because of the need to distinguish different types of addressing techniques, we also give each a separate name. The method of putting data immediately below the instruction and fetching the value by making use of the PC is called *immediate addressing*, and putting an address below the instruction and fetching the data using the above described two-step process is called *absolute addressing* (because the programmer must know the exact address of the data) or *direct addressing* (because the instruction indicates directly where to find the data).

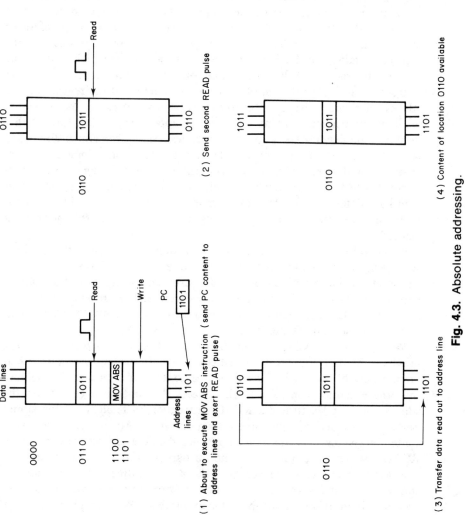

Fig. 4.3. Absolute addressing.

We now introduce the idea of *indirect addressing*. Suppose we have a program which refers repeatedly to a particular item of data. Obviously, we can put its address below every instruction that uses it. However, this is not a very efficient way of utilizing memory space. Instead, only one instruction is contained in the early part of the program which loads the address into a register, and all the instructions which use the data would contain a reference to this register. The addressing mode specification now indicates that the register contains, not the data item itself, but the address which contains the relevant instruction. When the execution control unit receives such an instruction, it fetches the data by placing the content of the register on the address lines and activating a READ instruction which causes the content of the memory location containing the data to appear on the data lines so that the execution may use the value. Indirect addressing is sometimes also called *deferred addressing*, since there is a delay in getting the data. (Some computers have double-deferred addressing, with the register pointing to a memory location, which itself contains the address of the data. In theory one could even have deferments of three or more levels, though it is hard to see what practical value they would offer.) Figure 4.4 illustrates this addressing method.

It should be pointed out that the PC does not itself play any part in indirect addressing. However, the instruction which sets up the address in the register has to make use of the PC; this instruction would have the address immediately below itself, and during the execution the processor would use the PC to fetch the address and put it into the register. Thus, the PC plays an indirect role in indirect addressing.

Yet another addressing method making use of the PC is *indexed addressing*, in which the data address has to be computed, by adding the value stored immediately below the instruction to the content of a register. To achieve this, the execution control unit first places the content of the PC on the address lines and activates a READ instruction to fetch the value stored below the instruction. This value is then sent to the Adder together with the content of the register, and their sum is placed on the address lines and a new READ instruction activated to obtain the data required. Indexed addressing is a useful technique whenever we make use of addresses which contain a constant part and a variable part. For example, we may wish to process successively individual items in a table of numbers. We then place the index in a register, and place the starting address below the instruction which refers to this register and has an addressing mode specifying indexed addressing. As the value in the register increases, the same instruction would then access successive elements of the table: for example, if the register contains 3, the sum of the two parts would yield the address of the 3rd item of the table, i.e. starting address $+3$, and increasing the register to contain 4 would make the sum starting address $+4$, which would refer to the 4th item.

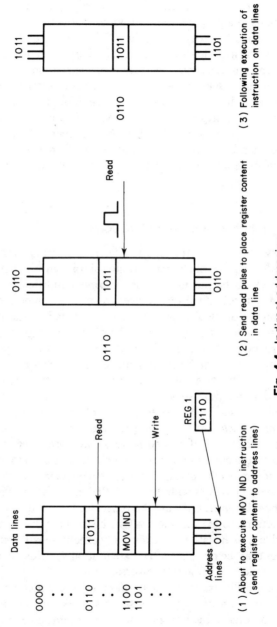

Fig. 4.4. Indirect addressing.

(1) About to execute MOV IND instruction (send register content to address lines)

(2) Send read pulse to place register content in data line

(3) Following execution of instruction on data lines

Having discussed the role of the program counter in data access, we now turn to its role in program control. Earlier we stated that the individual instructions in a program (with or without attached constants and addresses) will be executed in sequence if the PC is incremented by one each time it is used in an instruction or data fetch. There is, however, also the need to provide for non-sequential execution of instructions. For example, we may wish to execute selectively one out of several alternative segments of code depending on the results of earlier computation, or to execute the same set of instructions a multiple number of times. The machine instructions designed for these purposes are called *branches* or *jumps*. However, because these instructions are related to the condition code register we need to consider now how these operate in terms of the state of this register.

4.2.2 Branches and Condition Codes

We know that each instruction resides in a memory location, and therefore has an address. A branch instruction is simply an instruction which causes a new address to be put into the PC if some condition is satisfied. Since the execution control unit always fetches the next-executed instruction according to the content of the PC, putting a new address into the PC ensures that the normal sequence of execution will be disrupted, and instead of going on to the instruction below, execution will start in a different part of the program. So we say that the processor has made a jump or a branch.

Figure 4.5 illustrates the use of branch instructions in the implementation of a common program structure, if IF . . . THEN . . . ELSE . . . construct. The first instruction compares two numbers x and y, and the second is a *conditional branch*, ordering the processor to execute the program segment starting at address B if x is equal to y. Suppose, however, x and y are not equal, then the condition for the branch has not been met, and therefore the branch does not take place. Hence, the next executed instruction is the one immediately below, at address A. At the end of code segment A there is an *unconditional branch*, ordering the processor to skip over code segment B and start executing code from address C. This ensures that, regardless of whether alternative A or B was taken, eventually execution will proceed from C. Also, it ensures that the two alternatives are mutually exclusive, i.e. if A is executed then B would not be, and vice versa.

Figure 4.6 illustrates another common program construct, the *loop*. The code shown is meant to process successive elements of a table. It starts by putting an initial value into a register. Every time an element of the table has been processed this number is increased by 1 and compared with the upper limit MAX. If this limit has not yet been reached, a jump is made back to the

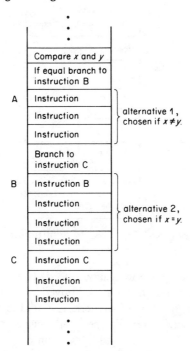

Fig. 4.5. The IF . . . THEN . . . ELSE construct

start of the loop to repeat the processing on a new element. If the limit has been reached, the branch does not take effect and the instruction below is executed, which means that the processor has completed its repetitive operations and continues with further processing in the program.

Two points are relevant here. First, since a branch instruction alters the sequence of execution by putting a new address into the PC, the programmer must place some addressing information below the branch. This does not, however, have to be the complete address, and on some machines the processor assumes that the *difference* between the new address and the present value of the PC is stored, and the execution control unit would fetch a value from beneath the branch instruction, add this to the current value of the PC, and then start a new instruction fetch using the changed value of the PC. A positive added value would cause a jump forward, and a negative value a jump backward. This technique, called *relative addressing*, has two merits. First, it permits us to store only a byte below the jump, rather than a full 16-bit address, provided that the instruction we wish to jump to is not too far away from the current location so that the difference between the new and old addresses is in the range $\pm 2^7$. This is adequate for most purposes, though

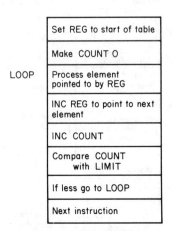

Fig. 4.6. The LOOP construction.

occasionally a special *long jump* instruction needs to be invoked. The second advantage is that when the program is moved to somewhere else in the memory, the addresses of all the instructions would change but differences between pairs of instructions remain the same since all the addresses change by the same amount. Hence, the addressing information stored with the branches remains correct and need not be modified.

The second relevant point is that there must be some way of recording the results of a compare instruction in the processor, to make them available to the branch instruction executed immediately after the compare. In fact, it is normal to record not just the results of compare instructions, but virtually all logical and arithmetic operations. For example, it is possible to write code of the type

> add x to y
> if positive branch to address D

The processor, after executing the ADD instruction, would record information about the result, e.g. whether it was zero, whether it was positive or negative, whether it had even or odd parity, whether there was an overflow (whether the result was too large to be correctly stored in the machine). Indeed, the "compare x with y" instruction is nothing more than a subtract and then recording of the conditions created—the actual arithmetic computed is discarded since this is not required. In contrast the instruction, "subtract y from x" causes the computed result to be retained in the location previously storing x as well as recording the information about the result. Similarly, the instruction, "negate y" would replace y in the location previously containing y in addition to recording information about y, whereas "test y" would merely record information about y without altering it.

The condition information is stored in the condition code register, CC of Fig. 4.1(*a*). The figure shows that this register is connected to the output path of the arithmetic unit, so that whatever result is produced by the unit would change the CC. Now this means we cannot carry out any operations between a compare and a branch instruction that has to make use of the results of the compare, since the CC may be altered by the results of the compare operation and the branch would no longer reflect the conditions it is meant to use. At the same time, we see that if we are able to save the value of the CC, and later restore it, then the branch would reflect the designated conditions even though some other processing might have occurred between the save and the restore operation. Indeed, it is generally true that if we save the contents of the memory (both program and data) and the registers, and load some completely new values into the processor to perform work completely unrelated to the previous program, the program would recommence as if the interruption had not occurred. A change of processing like this is called a *context switching*. Such switching is required constantly in the sort of multi-programming and multi-tasking environment discussed in Chapter 1. Provided that each task does not violate the memory space used by other tasks and does not alter anything which does not belong to it, many such context switchings can occur in the course of a period of processor operation.

As mentioned earlier, a typical microprocessor would have a condition code register capable of recording the zero/nonzero, negative, overflow/no overflow, even/odd parity, and carry/no carry properties of the latest computed result, and we can have branch instructions which make use of individual parts of the CC to decide whether a branch is to occur. There can, however also be branches which are determined by some logical combination of the individual bits of the CC. For example, one can have a "branch if non-positive" instruction, which causes a jump if either the negative bit or the zero bit is turned ON. This is a more obvious case: other types of branches exist whose relation with the CC bits are not so obvious and numerous examples will be found in the literature on computer systems design.

4.2.3 The Stack Pointer

The *stack* is an extremely important concept in computing, with wide applications in many areas, some to be discussed later in this book. For the time being, we shall introduce the concept by discussing its connection with *subroutine jumps*. A subroutine is a program segment which resides in a separate location from the main program, and is meant to perform some function which is required a number of times at different places in the main program. We could of course provide this function by including the code segment in the main program where it is required. However, this would again be wasteful of program memory, and it is better if we have the code segment in a separate part of the memory, but at an address known to the main program, which makes a jump to the subroutine to perform the required function. At the completion of the subroutine a *return* is made to the main program where it was previously interrupted by the subroutine jump, and execution continues in the main program, until it encounters some other subroutine jump when another transfer to subroutine and return from subroutine would occur.

While both the subroutine jump and subroutine return cause a transfer of program execution from one instruction sequence to another, they are quite different from normal branch instructions in one important aspect; while branch instructions transfer control to specific addresses provided with the branches, a subroutine return does not have a specific address to transfer to; the main program can call a subroutine at several different program locations and returns have to be made to these locations. Thus at the time the programmer writes the subroutine he does not know the reutrn address. Instead, the address is provided by the main program *at the time the subroutine is called*. That is to say, when the main program makes a subroutine jump, the processor automatically puts the return address in a place accessible to the subroutine. When the subroutine is completed and a return is made, the processor fetches the return address from its storage location and makes a jump to that location to resume the main program.

The question then arises: where should the return address be placed during the execution of a subroutine? It is possible to reserve a register or a standard memory location-for this purpose, but unfortunately this can cause difficulties. Consider what happens if the subroutine itself contains its own subroutines. The subroutine, when making its own subroutine jump, would put a return address into the same register or memory location which contains the return address put there by the main program. This second return address would therefore erase the first one, and a return to the main programme can no longer be made!

We see that to handle a situation like this correctly, we must reserve a series of registers or memory locations, where we can locate the new return address next to the first address. If the second subroutine itself calls yet a third subroutine, then we need to have space for three return addresses, and for a chain of N subroutine calls and returns we would require N consecutive registers or memory locations. To illustrate this, Fig. 4.7 shows a chain of 3 subroutine calls and the return address handling. When program A calls subprogram B, the processor puts return address A at the bottom of memory block reserved for this purpose. When program B calls C, the return address B is placed in the location above that containing address A; and when program C calls program D, the return address C goes above address B. when subroutine D finishes, the processor would take the address C from the list

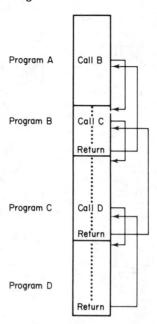

Fig. 4.7. Chained subroutine calls.

and use it to return to program C, which also completes in due course, whereupon the address B is taken from the list, and later the address A is used to return to the main program. By then the list of addresses would be empty (Fig. 4.8).

It is important to note two fundamental properties of the list of return addresses; it is a *dynamic vector* in the sense that it is a multiple element vector whose size varies as execution proceeds; also, it has a *last-in-first-out* structure, in that the last address added to the list, C, is the first one to be removed, As we mentioned earlier, such a structure appears frequently in computing, and hence it is given the special name of stack (elements are "stacked" on top of each other).

We are now at last able to discuss the register stack pointer provided in the Intel 8080 processor. In this processor the stack is simply some part of the data memory reserved by the programmer for the storage of return addresses and other information using this last-in-first-out structure. The stack pointer always contains the address of the top element of the stack. Whenever the processor executes a subroutine jump, it automatically puts the return address into the memory location above the previous top element of the stack, and also decreases the value of the stack pointer so that it contains the address of the new top of the stack, e.g. if the previous top of stack was at

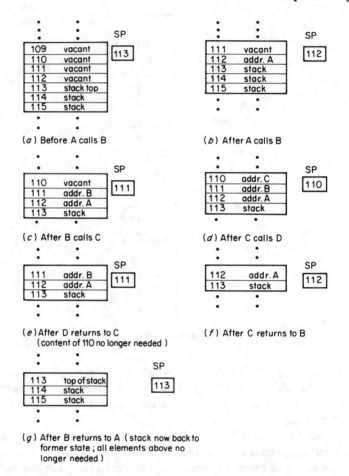

Fig. 4.8. Stack growth and contraction during chained calls/returns.

address 1016, the execution control unit would put the return address (16 bits long) into the memory locations 1014/5, and at the same time change the content of the stack pointer from 1016 to 1014. At any time the program can read the value of the top of stack by making use of the stack pointer register in indirect mode. Also, when the processor executes a subroutine return instruction, it merely reads the top of stack, and puts the value read out into the PC, which ensures that the next instruction fetched comes from the return address in the program which originally called the subroutine. It also increases the SP to point to the location below.

Before we close this subsection it is useful to discuss the concept of "return address". We know that when the processor is executing an instruction the

PC contains the address of the next instruction, below the current one. This is true for any instruction, including a subroutine jump, which, however, disrupts the sequence of execution by putting the address of the subroutine into the PC so that the next instruction executed is in the subroutine. When a return to the main program is executed, the processor needs to restore the value of the PC to that existing just before the subroutine jump, so that it will once again refer to the instruction below the subroutine jump, and the execution of the main program will resume from there. The required result will be achieved if during a subroutine jump the processor:

(a) puts the current content of the PC on the stack;
(b) decreases the content of the SP to point to the location above the previous top of stack;
(c) puts the address of the subroutine into the PC; and during a subroutine return it;
(d) puts the top of stack (which contains the old value of the PC, being the address of the instruction below the subroutine jump) back into the PC;
(e) increases the content of the SP to make it point to the element below.

In short, the execution of the subroutine jump and return instructions involves merely the transfer of values between the PC and the stack, and increasing or decreasing the SP. In contrast, ordinary branch instructions simply cause new values to go into the PC, without involving the stack in any way.

Whereas in the Intel 8080 the stack is simply a part of memory reserved by the programmer for this purpose, other arrangements are possible. In the earlier 8008 processor, for example, the stack is a set of seven registers. Successive subroutine jumps would cause values of the PC to be moved into these seven registers, and subroutine returns would cause values to be returned from there to the PC. The limited number of registers, however, makes long chains of subroutine calls impossible, as well as precluding the use of the stack for saving information other than return addresses. With a stack in the memory pointed to by an SP this limit is removed, which is one of the reasons why 8080 is more powerful than 8008. Another useful thing to mention is that, putting new items onto the stack is usually called *pushing*, and taking the top elements off the stack is called *popping*.

4.3   INTEL 8080 MACHINE INSTRUCTIONS

We now proceed to study in more detail the machine instruction set of the Intel 8080 microprocessor. The processor design provides for the execution of over a hundred different instructions. We shall consider them under the following five categories.

(a) *Data movement instructions*. These instructions cause the movement of items of information between registers, or between a register and a memory location. The total number of such instructions is quite considerable, because of the variety of addressing modes, as well as constraints on instruction formats due to the short wordlength of the processor, which forces the designer to implement a more complex design than would have been necessary otherwise. (We shall discuss this problem in a wider context in Section 4.6.)

(b) *Arithmetic and logical instructions*. All the instructions which combine two items of data arithmetically or logically, or perform operations on one number, are included here. On the Intel 8080 such instructions always involve the register A, also known as the accumulator.

(c) *Jump instructions*. These include branches, subroutine jumps, and subroutine returns.

(d) *Pointer manipulation instructions*. These instructions make useful changes on addresses stored in registers. Prominent are instructions involving the stack pointer, but there are also those which operate on some of the general purpose registers.

(e) *Miscellaneous*. The remaining instructions cannot be neatly characterized by a common description, and are classified under this heading.

We start our discussion with the general format of Intel 8080 instructions. Since the majority of instructions order the processor to perform some operation on some data, the instruction format must include information about *what* operation to perform on *which* data. Thus, in general an instruction has an *opcode* and one or more *address fields*, depending on whether the operation involves one or two items of data. Some of the instructions may also have attached data: immediate mode instructions would have data below them, and direct mode (absolute mode) instructions would have addresses below them, while input/output instructions would have device numbers below. The total size of an instruction is thus variable. One without any attached information would occupy just a byte, one with a byte of data or an I/O device number occupies two bytes, one with an address would occupy three. However, the opcode is always in the first byte, together with addressing mode/register information, laid out according to the following general form

OPCODE	DEST	SOURCE
|	| |	| |

bits	7	6	5	4	3	2	1	0

We see that the opcode is only two bits long, which would only allow four different instructions. However, a far greater number of instructions is provided by making use of the DEST (destination) and SOURCE fields as extensions to the opcode whenever they are not required for their normally intended functions, as we shall explain.

4.3.1 Data Movement Instructions

We start with the MOV instruction, whose function is to copy the content of one register to another, and whose format is 01dddsss, where ddd and sss may each be any 3-bit binary number other than 6(110). The value sss specifies the register whose contents are to be copied and ddd the register to receive this information, i.e. the source and destination register respectively. For convenience in register identification within the instruction format, the registers are given numbers where E = 0, C = 1, D = 2, E = 3, H = 4, and L = 5 (A is identified separately as a special accumulator register and made equal to 7). Thus, the string 01001100 means "copy the content of register H into register C". The previous content of the destination is erased by this instruction, but the source retains its value.

What if either ddd or sss equals 110? As remarked earlier, this does not refer to any of the seven registers. Instead it means that the destination or the source is a memory location, whose address is in the registers H and L. For example, suppose register H already contains 00110001 and L contains 10011010, then the instruction 01110111 means "copy the content of register A into the memory location with address 0011000110011010", while the instruction 01111110 would mean "copy the content of memory location 0011000110011010 into register A". In short, the DEST or SOURCE field indicates *indirect* addressing via the address pointer H-L whenever its value is 6, whereas the values 0 to 5 and 7 indicates *register* addressing.

Before executing a MOV instruction with addressing code 6 we must first ensure that the registers H and L contain the correct 16-bit address. As discussed in Subsection 4.2.1, this requires an instruction which moves a constant stored below, i.e. an *immediate mode* move instruction. On the Intel 8080 several such instructions are provided. First there is the MVI instruction, which copies *one* byte into *any* of the seven general registers, including of course either H or L. The instruction has the format 00dd110. The 00 in the first two bits and the 110 in the last three bits *together* specify an MVI instruction. If either part is different then this would be some other instruction altogether. The DEST field, on the other hand, has the same meaning as with the MOV instruction, that is, if it is 0,1, . . . , 5 or 7 the MVI instruction would copy a constant into the registers B, C, . . . , L or A. If DEST=6 then the constant is copied into the memory location specified by the value of the H-L registers. To illustrate, the instruction/data combination

00111110/00000001 would copy the number 00000001 into register A, while the combination 00110110/00000001 would cause 00000001 to go into a memory location, provided that we have previously put its address into the registers H and L. The following short code segment would first load the address, and then use the address to put a constant into the specified memory location, after first reading out its content into register A.

00100110	(copy the byte below into register H)
00110001	(constant—first part of address)
00101110	(copy the byte below into register L)
10011010	(constant—second part of address)
01111110	(copy the content of memory location pointed to by registers H and L into register A, i.e., location 0011000110011010)
00110110	(copy the byte below into memory location pointed by registers H and L, i.e., put 00000001 into location 0011000110011010)
00000001	

In the above code segment two instructions were executed to load two halves of the address 0011000110011010 into the H and L registers. The Intel 8080 design also provides the quicker method of a double load instruction. The instruction 00100001 asks the processor to load the two bytes below the instruction into registers H and L. Thus, the first four bytes of the above code segment may be simplified into three bytes.

00100001	(copy the two bytes below into H and L)
00110001	(first part of address, going into H)
10011010	(second part of address, going into L)

Similarly, the instruction 00010001 loads two bytes into the D and E registers, while 00000001 loads two bytes into B and C. These three instructions are called Load Immediate instructions, and designated LXI (where X stands for index register). We should note that only these three configurations of paired registers are permissible. It is not possible to ask for two bytes to be loaded into, say, registers C and D by means of an LXI instructions nor is it possible to pair register A with any other.

Other instructions exist which single out the register A for special treatment. We know that the MOV instruction permits the movement of information between any register and a memory location pointed to by the H and L registers. With register A, however, it is also possible to move data between it and a memory location whose address is in either B/C or D/E, by means of the STAX and LDAX instructions. The former *stores* the content of A, e.g. STAX B would copy the content of A into the memory location whose address is in B/C, and the latter *loads* data into A, e.g. LDAX D would copy the content of the memory location pointed to by D/E into A.

In all the above instructions a memory location is accessed only after we have first set up its address in a pair of registers. This is efficient if we plan to use the same address a number of times, but if it is used only once then it would be desirable to employ some method which avoids designating a pair of registers in this way. The method of absolute addressing or direct addressing described in Section 4.2.1 achieves this, but unfortunately its availability in the Intel 8080 is rather limited. The STA (00110010) instruction stores the content of A into a memory location whose address appears below the instruction, e.g.

> 00110010
> 00110001
> 10011010

requests the processor to store the content of A into the memory location 0011000110011010. The LDA instruction (00111010) would move the content of a memory location into A. SHLD (00100010) and LHLD (00101010) do the equivalent for the register pair H/L, but would move two bytes rather than one, e.g.

> 00101010
> 00110001
> 10011010

would take the contents of H and L and put them into the memory locations 0011000110011010 and 0011000110011011.

It is of interest to ask why the Intel designers did not make similar instructions available for other registers. The reason lies in the lack of instruction formats. Given an 8-bit format, one can only define so many different operations, registers and addressing modes. The MOV instruction alone uses 63 of the 256 available instruction formats, and even with the imposed restrictions on registers available with other data movement instructions, they still use up nearly half the available instruction formats. The instruction format limitation appears in one form or another with all 8-bit microprocessors, and we shall discuss this further in Section 4.6.

4.3.2 Jump Instructions

The intel 8080 has three sets of jump instructions—branches, subroutine calls and subroutine returns. As discussed in Section 4.2.2, branch instructions have addresses stored below themselves, subroutine calls are similar to branches, except that they also cause return addresses to be saved on the stack, whereas subroutine returns do not have attached addresses, but derive return addresses from the stack. Each of the three jumps may be unconditional, or be conditional upon one of the bits in the CC register. For example, the instructions JC (jump-on-carry), CC (call-on-carry) and RC (return-on-carry) will cause a jump if and only if the carry bit of the CC is 1, i.e. if the previous instruction caused an arithmetic carry. On the other hand, the instructions JNC (jump on no carry), CNC and RNC cause a jump if and only if the carry bit is 0. Other jump instructions cause program transfers depending on whether the zero bit, sign bit or parity bit is 0 or 1. In addition, there are the JMP, CALL and RET instructions which will cause a jump regardless of the content of the CC register. Finally, the instruction PCHL causes the content of the H and L registers to be copied to the PC. This causes a jump to the memory location pointed to by H/L since its address is now in the PC. It is thus an unconditional branch with indirect addressing mode, the assumption being that the address to jump to has been previously set up in the H/L registers.

4.3.3 Accumulator Manipulation Instructions

The format of these instructions is rather similar to that of the MOV instruction, being 10dddsss. However, whereas with MOV the data may be copied *from* any register *to* any register, with the present instructions the destination has to be register A, or the accumulator. Thus, the three bits ddd do not specify the destination, but specify one out of eight possible arithmetic or logical operations on register A. Thus, 10000sss cause a number to be added, while 10001sss cause a number to be subtracted, while 10100sss cause an AND operation. In all such operations register A is a participant, and the result of the operation is always placed in A, whose previous content would be erased. The second participant is specified by the value of sss. As in the MOV instruction, sss = 0,1, . . . , 5 or 7 means the number comes from the registers B,C, . . . , L or A. Thus, the instruction 10000111 adds the content of register A into itself, i.e. doubles the value of A. Instruction 10010111 subtracts A from itself, which produces a 0. (One could also cause a 0 to appear in A using the instruction MVI A, 00000000, but the method just described is faster and one byte shorter.) Instruction 10100111 would AND A with itself, and thus leave it unchanged. This instruction is useful however since, as discussed in Section 4.2.2, the content of the CC register always reflects the result of the latest instruction. Since the latest instruction simply reproduced the content of A, the CC register thus reflects the value of A, i.e. if the content of A is 0 then the zero bit is set, and if the value is negative then the negative bit is set, etc. Hence, the instruction 10100111 may be followed by a conditional branch instruction, which would cause a jump if the value of A fulfills some specified condition.

Table 4.2. Intel 8080 Instruction set: summary of processor instructions

Mnemonic	Description	D_7	D_6	D_5	D_4	D_3	D_2	D_1	D_0	Clock Cycles
Move, load and store										
MOVr1r2	move register to register	0	1	D	D	D	S	S	S	5
MOV M r	move register to memory	0	1	1	1	0	S	S	S	7
MOV r M	move memory to register	0	1	D	D	D	1	1	0	7
MVI r	move immediate register	0	0	D	D	D	1	1	0	7
MVI M	move immediate memory	0	0	1	1	0	1	1	0	10
LXI B	load immediate register pair B & C	0	0	0	0	0	0	0	1	10
LXI D	load immediate register pair D & E	0	0	0	1	0	0	0	1	10
LXI H	load immediate register pair H & L	0	0	1	0	0	0	0	1	10
STAX B	store A indirect	0	0	0	0	0	0	1	0	7
STAX D	store A indirect	0	0	0	1	0	0	1	0	7
LDAX B	load A indirect	0	0	0	0	1	0	1	0	7
LDAX D	load A indirect	0	0	0	1	1	0	1	0	7
STA	store A direct	0	0	1	1	0	0	1	0	13
LDA	load A direct	0	0	1	1	1	0	1	0	13
SHLD	store H & L direct	0	0	1	0	0	0	1	0	16
LHLD	load H & L direct	0	0	1	0	1	0	1	0	16
XCHG	exchange D & E H & L registers	1	1	1	0	1	0	1	1	4
Stack ops										
PUSH B	push register Pair B & C on stack	1	1	0	0	0	1	0	1	11
PUSH D	push register Pair D & E on stack	1	1	0	1	0	1	0	1	11
PUSH H	push register Pair H & L on stack	1	1	1	0	0	1	0	1	11
PSH PSW	push A and Flags on stack	1	1	1	1	0	1	0	1	11
POP B	pop register Pair B & C off stack	1	1	0	0	0	0	0	1	10
POP D	pop register Pair D & E off stack	1	1	0	1	0	0	0	1	10
POP H	pop register Pair H & L off stack	1	1	1	0	0	0	0	1	10
POP PSW	pop A and Flags off stack	1	1	1	1	0	0	0	1	10
XTHL	exchange top of stack H & L	1	1	1	0	0	0	1	1	18
SPHL	H & L to stack pointer	1	1	1	1	1	0	0	1	5
LXI SP	load immediate stack pointer	0	0	1	1	0	0	0	1	10

(cont.)

Table 4.2. (*cont.*) Intel 8080 Instruction set: summary of processor instructions

Mnemonic	Description	D_7	D_6	D_5	D_4	D_3	D_2	D_1	D_0	Clock Cycles
INX SP	increment stack pointer	0	0	1	1	0	0	1	1	5
DCX SP	decrement stack pointer	0	0	1	1	1	0	1	1	5
Jump										
MP	jump unconditional	1	1	0	0	0	0	1	1	10
JC	jump on carry	1	1	0	1	1	0	1	0	10
JNC	jump on no carry	1	1	0	1	0	0	1	0	10
JZ	jump on zero	1	1	0	0	1	0	1	0	10
JNZ	jump on no zero	1	1	0	0	0	0	1	0	10
JP	jump on positive	1	1	1	1	0	0	1	0	10
JM	jump on minus	1	1	1	1	1	0	1	0	10
JPE	jump on parity even	1	1	1	0	1	0	1	0	10
JPO	jump on parity odd	1	1	1	0	0	0	1	1	10
PCHL	H & L to program counter	1	1	1	0	1	0	0	1	5
Call										
CALL	call unconditional	1	1	0	0	1	1	0	1	17
CC	call on carry	1	1	0	1	1	1	0	0	11/17
CNC	call on no carry	1	1	0	1	0	1	0	0	11/17
CZ	call on zero	1	1	0	0	1	1	0	0	11/17
CNZ	call on no zero	1	1	0	0	0	1	0	0	11/17
CP	call on positive	1	1	1	1	0	1	0	0	11/17
CM	call on minus	1	1	1	1	1	1	0	0	11/17
CPE	call on parity even	1	1	1	0	1	1	0	0	11/17
CPO	call on parity odd	1	1	1	0	0	1	0	0	11/17
Return										
RET	return	1	1	0	0	1	0	0	1	10
RC	return on carry	1	1	0	1	1	0	0	0	5/11
RNC	return on no carry	1	1	0	1	0	0	0	0	5/11
RZ	return on zero	1	1	0	0	1	0	0	0	5/11
RNZ	return on no zero	1	1	0	0	0	0	0	0	5/11
RP	return on positive	1	1	1	1	0	0	0	0	5/11
RM	return on minus	1	1	1	1	1	0	0	0	5/11
RPE	return on parity even	1	1	1	0	1	0	0	0	5/11
RPO	return on parity odd	1	1	1	0	0	0	0	0	5/11
Restart										
RST	restart	1	1	A	A	A	1	1	1	11
Increment and decrement										
INR	increment register	0	0	D	D	D	1	0	0	5
DCR	decrement register	0	0	D	D	D	1	0	1	5

Mnemonic	Description	Instruction code								Clock Cycles
		D_7	D_6	D_5	D_4	D_3	D_2	D_1	D_0	
INR M	increment memory	0	0	1	1	0	1	0	0	10
DCR M	decrement memory	0	0	1	1	0	1	0	1	10
INX B	increment B & C registers	0	0	0	0	0	0	1	1	5
INX D	increment D & E registers	0	0	0	1	0	0	1	1	5
INX H	increment H & L registers	0	0	1	0	0	0	1	1	5
DCX B	decrement B & C	0	0	0	0	1	0	1	1	5
DCX D	decement D & E	0	0	0	1	1	0	1	1	5
DCX H	decement H & L	0	0	1	0	1	0	1	1	5
Add										
ADD	add register to A	1	0	0	0	0	S	S	S	4
ADC	add register to A with carry	1	0	0	0	1	S	S	S	4
ADD M	add memory to A	1	0	0	0	0	1	1	0	7
ADC M	add memory to A with carry	1	0	0	0	1	1	1	0	7
ADI	add immediate to A	1	1	0	0	0	1	1	0	7
ACI	add immediate to A with carry	1	1	0	0	1	1	1	0	7
DAD B	add B & C to H & L	0	0	0	0	1	0	0	1	10
DAD D	add D & E to H & L	0	0	0	1	1	0	0	1	10
DAD H	add H & L to H & L	0	0	1	0	1	0	0	1	10
DAD SP	add stack pointer to H & L	0	0	1	1	1	0	0	1	10

Again as with the MOV instruction, if the value of sss is 110 then the second participant in the operation comes from a memory location, whose address is in H/L. Thus, the instruction 10000110 adds a number read out from memory to register A. Also, in analogy to the MVI instruction there is a set of immediate mode instructions, which combines a byte stored below the instruction with the content of register A. These instructions have the format 11ddd110, in which ddd again has the function of specifying which operation is required. Thus, the combination 11000110/00000001 adds 1 to the accumulator, while 11010110/00000001 subtracts 1.

As may be seen from the complete list of Intel 8080 instructions given in this chapter (Table 4.2), there are actually two add instructions (ADD, ADC) and two subtracts (SUB, SBB). The ADC instruction really adds three

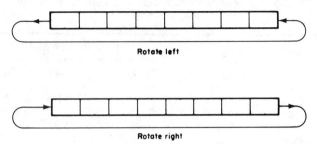

Rotate left

Rotate right

Fig. 4.9. Rotate instructions.

numbers together, the accumulator, the second participant as specified by sss, and the carry bit of the CC register (the carry bit, being either 0 or 1, is a one-bit binary number). We remember that when two numbers are added any carry out of the most significant digit is recorded in the CC register. The ADC instruction then allows us to add this is "excess value" to the next stage of the arithmetic process. The SBB instruction serves a similar function, allowing the "deficit" of the previous stage of subtraction to be taken from the next stage, again by making use of the carry bit of the CC, since with subtractions the borrow, rather than carry, is the relevant information recorded in the CC.

Finally we have the four instructions RLC, RRC, RAL and RAR which rotate the content of register A by one bit, either left or right. The former two simply rotate the eight bits in A, with the bit shifted out at one end coming back into A at the other end (Fig. 4.9), but the latter two include the carry bit of the CC in the rotation, and the bit shifted out of register A will occupy the carry bit position in the CC, while the former carry bit will get shifted into the vacated end of A (Fig. 4.10). This extended shift has several uses. As mentioned earlier, the carry bit represents excess from one stage of a multi-stage addition or deficit from a subtraction. The RAL and RAR operations

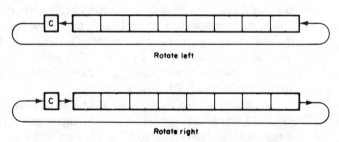

Rotate left

Rotate right

Fig. 4.10. Rotate-with-carry instructions.

permit the value to be moved into register A and become part of a number. Also, in conjunction with instructions which force the carry bit to be zero (two instructions STC and CMC, which belong to Subsection 4.3.5, affect the carry bit) the RAL and RAR cause a bit to be moved out of register A without coming back in at the other end, which is useful for extracting part of the content and removing the part we do not need. Two more accumulator manipulation instructions are CMA, complement A, which turns the zeros in A into ones and vice versa, and DAA, decimal adjust A, which, used after an addition, changes the result to what would have been produced if the two halves of each number are 4-bit BCD digits.

4.3.4 Register Manipulations

Under this heading are a group of instructions for increasing or decreasing register values by 1, a number of instructions involving the stack pointer, and a few other instructions on registers. The INR instruction (00ddd110) increments a register identified by the number ddd (if ddd = 6 the memory location pointed to by H/L is incremented), and the DCR instruction (00ddd101) decrements a register or memory location. The INX and DCX instructions increment or decrement the register-pairs B/C, D/E or H/L, and are much used in table accessing. By loading the starting address of a table of numbers into a pair of registers and incrementing the address successively, we can fetch sequentially individual values of the table. Among the second group, the PUSH instructions place the register contents on top of the stack, two at a time. Thus, PUSH B (11000001) puts the contents of B and C on the stack, at the same time decreasing the SP by 2 to make it point to the new top of stack. PUSH D (11010001) and PUSH H (11100001) carry out the equivalent with D/E and H/L. PUSH PSW (11110001) saves the contents of register A and the CC. The instruction POP reverses the above operations, putting the top two elements of the stack into a pair of registers and at the same time increasing the value of the SP by 2 to point to the new top-of-stack. The instructions INX SP and DCX SP increase or decrease the SP by 1 without transferring data to or from the stack, while LXI SP puts a new 16-bit value, from below the instruction, into the SP, giving the programmer the means of creating a stack at an address chosen by himself (stack initialization). The instruction XTHL exchanges the two top bytes of the stack with the contents of H and L, without altering the height of the stack (or the SP), while SPHL moves the value in H/L into the SP, another way of establishing a new stack at a programmer-selected address. Finally, the DAD instruction takes a 16-bit value from either B/C, D/E, H/L or the SP and adds this to the content of H/L, and XCHG exchanges the content of the two register pairs D/E and H/L.

4.3.5 Miscellaneous

Here are included the I/O instructions IN and OUT, which specify a data transfer between the processor and an I/O channel, whose number is given below the instruction; the EI and DI instructions which permit (enable) or disallow interrupts, which we consider further in the next chapter; the NOP, no-operation instruction which does not initiate any action but is inserted in a program to adjust timing or sequence of instructions; the HLT, halt instruction; several instructions to change the carry bit of CC; and RST, restart instruction, a special jump instruction used in conjunction with interrupt routines.

4.4 ASSEMBLY PROGRAMMING

We saw in the preceding section that a machine program is made up of strings of ones and zeros in accordance with a defined instruction format. Every individual bit has to be precisely defined in order to produce a program that will execute correctly. To write a machine program directly the programmer will not only have to remember the opcode for every instruction and the restrictions on available addressing modes and registers, he also has to insert the correct address for every data item used in his program at the right place. Such a process is extremely tedious and error-prone, and implementing even a fairly simple program is a difficult and time-consuming process. A few years ago when microprocessors were in an early stage of development (and also when computers were first invented earlier in the century) programs were in fact written this way. Today, however, the programmer can make use of a number of facilities to simplify his task and develop reliable programs of fairly high complexity. One such facility is the *assembler*, which produces the binary machine program from a source program written in a more manageable notation called the *assembly language*, a language which is closely related to the instruction set of the processor and can be translated into the machine program using a well defined algorithm.

A basic assumption of assembly programming is that names composed of letters are more meaningful than numbers and hence easier to remember without error. The language therefore assigns a name to each instruction, register, and every important memory location, whether occupied by an instruction or an item of data. Thus, instead of 01111000 the programmer writes MOV A,B if he wants to copy the value in register B to register A, and the assembler translates the latter into the former, or in computer jargon, produce the object code from the source code. Similarly, if the programmer writes MVI A, 1 followed by STAX B the assembler produces from these the three bytes

$$00111110$$
$$00000001$$
$$00000010$$

The identification of each register or instruction with a fixed string of ones and zeros is a fairly simple task. Giving names to memory locations and identifying them with addresses is more complex and usually requires a *two-pass assembly* process, which we discuss below together with other features of assembly languages.

4.4.1 Pseudo-code and Two-pass Assemblers

In addition to instructions like MOV A,B which are executable and are translated into machine instructions like 011111000, an assembly program also contains a number of pseudo-instructions or pseudo-ops whose function is to provide information to the assembler about the program, such as how many items of data are required by the program, the properties of each item, and where in the program the data should be placed. These are data-definition pseudo-ops. There are also pseudo-instructions which give names to code segments, constants, memory locations, etc., so that other parts of the program may refer to them. We also have pseudo-ops which tell the processor to handle the program or part of it in some specified fashion, such as putting it at a particular place in the memory, making the program available from some other program by defining special entrance points, or to designate some undefined addresses whose values are specified later by linking the program into another program that contain those addresses (*external references*).

For example, The Intel 8080 assembler permits data definitions along the following lines

```
        ORG  1000
CONST   EQU  63        ; THIS DEFINES A CONSTANT—
                         NO : NEEDED
INIT:   DW   0         ; DEFINES A TWO-BYTE VARIABLE
                         WITH INITIAL VALUE ZERO
ERMSG:  DB   "ERROR"   ; DEFINES A TEXT STRING FOR
                         ERROR MSG EACH LETTER TAKES
                         A BYTE
TABLE:  DS   100       ; ALLOCATES STORAGE FOR A
                         TABLE OF MAXIMUM SIZE 100
                         BYTES
```

Here ORG asks that the data being defined should be placed from location
1000 onward (origin); EQU says that the programmer may write CONST
instead of 63 in the program. This is useful if 63 appears many times but may
later be altered to some other value throughout the program. By using the
EQU definition and writing CONST instead of 63, the programmer only has
to alter the 63 in the definition to the new value without changing the rest of
the program.) DW asks the assembler to allocate a one-word memory location to the name INIT. Since this is the first memory location in the data
definition, the name INIT thus corresponds to the address 1000. The next
memory location reserved for data is then 1002, since the two bytes 1000 and
1001 have been allocated to INIT. Thus, ERMSG corresponds to the address
1002, and DB defines a string of five bytes, each containing a letter to make
up the alphabetical string "error". The next memory location available is
therefore 1007, which now has the name TABLE, and the DS statement
reserves the next 100 bytes for this table. Obviously, if further data definition
statements appear in the list then the addresses 1107 onward are allocated.

   With the above data definition the programmer is then able to write
instructions such as the following

STA	INIT	; STORES CONTENT OF A REGISTER INTO MEMORY LOCATION 1000
MOV	A, B	; MOVE CONTENT OF B TO A
STA	INIT + 1	; STORES CONTENT OF B (NOW IN A) INTO LOCATION 1001
LXI	B, TABLE	; PUT ADDRESS OF TABLE, WHICH IS BELOW THE INSTRUCTION, INTO B/C
LDAX	B	; LOAD MEMORY LOCATION POINTED TO BY B/C, i.e. TABLE, INTO A
ADI	CONST	; ADD 63 TO A

Not only will the assembler put the correct instruction values into the program, it also places the address of INIT below the first STA instruction, 1001
below the second STA, 1007 below the LXI, etc., further, it keeps count of
how many bytes of program are required. Thus, STA plus two bytes of
address takes 3 bytes, MOV takes only one, LXI, with a two-byte address
below, requires 3 bytes, but STAX takes only one, Hence, the above code
segment will altogether occupy 13 bytes of memory. If the program starts at
address 2000 (the program may ask for this using an ORG instruction at the
start of the code), then the next instructions added to the program will use
location 2013 onward. Thus, the assember keeps track of the address of each
new instruction in the program, so that if necessary jumps to each instruction
may be made. As with each item of data, the programmer may place a name
in front of each instruction together with a colon, and the assembler will

record the name and its corresponding address to facilitate a jump instruction. Suppose, for example, we add to the above code segment the following additional instructions

```
          SUB    A       ; SUBTRACT A FROM A TO MAKE A ZERO
LOOP:     STAX   B       ; PUT CONTENT OF A INTO LOCATION
          INX    B       ; POINTED TO BY B/C. THEN INC B/C
          INR    A       ; INCREASE A BY 1
          CPI    100     ; HAS A GONE UP TO 100 YET?
          JC     LOOP    ; IF NOT GO REPEAT
```

We see that the SUB instruction has address 2013, and therefore the name LOOP corresponds to address 2014. Hence the JC (jump if carry bit is 1 will have the address 2014 attached to it, and its execution would cause the instruction STAX to be re-executed.

Some explanation of the function of this code segment is useful here. We know that the previous LXI instruction placed the address of TABLE into B/C, and SUB made A contain 0. Hence, STAX B will put 0 into the memory location for TABLE. Now the address in B/C is increased, so it will point to the next memory location, and A also is increased to contain 1. This value is compared with the constant 100. The CPI operation (and CMP too) tries to subtract the second operand from the content of A. If the content of A is smaller than the constant, then a *borrow* would be generated. This is recorded in the *carry* bit of the CC. If the content of A is larger than or equal to the constant no borrow will be generated and carry bit will be zero. Hence, the JC instruction will cause a jump to LOOP if A has not reached 100. The STAX instruction will then put 1 into the location just below TABLE. The program repeats each time placing a larger number into the next memory location, until the numbers 0 to 99 have been placed into the 100 bytes allocated to the table. A would then have reached 100 and JC would no longer jump back to LOOP. Instead, the instruction below would be executed. We say that an *exit* from the loop occurs at the next part of the program.

Jump instructions may specify addresses that appear earlier or later in the program, and the programmer also has the freedom to put his data definitions almost anywhere in his program, whether above or below the instructions that make use of them. Thus, when the assembler encounters an instruction that refers to some address, the value of that address could very well be undefined as yet. This is why a program assembly is normally a two-pass process. The assembler first goes through the program instruction by instruction and data item by data item, counting the amount of memory used by each and assigning addresses to all the names. Gradually a symbol table is built up listing all the names encountered in the program a second time, filling in all the address values unavailable during the first pass. This makes the object program complete.

We have only provided the simplest discussion of the assembly process. Further details about the assembler facilities for each microprocessor may be found in the manufacturers' assembly programming manuals such as those listed in the reference section at the end of the chapter.

4.4.2 Macros

Earlier we discussed the concept of subroutines, which are self-contained segments of instructions required repeatedly in several parts of the program, and brought into use via subroutine calls and returns. Macros are meant to meet the needs for very short code segments. If such segments are implemented as subroutines, then the program will have to execute two jump instructions (call and return) only to execute a small number of instructions in the subroutine itself and, if the subroutine is called frequently, much time is wasted on the unproductive jump instruction. So instead such instruction segments are defined as macros, which the program may refer to, with the difference that each reference causes the assembler to insert the instructions given in the macrodefinition into the program. Thus, repeated macro references will simply cause multiple insertions of the same code at different places in the program. Whereas subroutines are meant to save space, macros are meant to save time.

For example, suppose in our program we wish to add frequently a number in a memory location to registers other than A. We would write the following macro definition

```
ADR          MACRO  REG
             PUSH   PSW
             MOV    A, REG
             ADD    M
             MOV    REG, A
             POP    PSW
             ENDM
```

Here the first statement is the macro definition pseudo-op, which permits subsequent calls of the macro by the name ADR (add to register), and ENDM signifies the end of the macro definition. Of the executable instructions the PUSH saves the content of A and the POP restores it, permitting the use of the register A in the interim. The first MOV instruction moves the value of the required register into A, and the ADD instruction adds the number in the memory location pointed to by H/L to A, after which the sum is moved back to the original register. Whenever our program wishes to make use of this macro, it will have the following macro call

```
ADR    B
```

which will cause the insertion of the five executable instructions from the macro definition into the program, with the register symbol REG replaced by B. The assembler will then be able to translate the instructions in the normal way, (Note: The macro should be used with care. Its effect is somewhat different from a normal ADD instruction because it cannot change the CC, which is restored to the state before the execution of the five instructions. Thus, it cannot be combined with the branch instructions. We can eliminate the PUSH and the POP, so that the CC will reflect the result of the addition, but the value of A would be affected by the macro execution, unless the program saves it before the macro call and restores it after. Here is an example of the limitation imposed by the instruction set, which will be further discussed later.)

In the last example, the macro ADR has the attached name REG, which will be changed to B by the macro call ADR B. The program using the macro may, by putting an alternative register symbol behind ADR, replace REG by some other register. REG is said to be a macro *parameter*, which is subject to specification by the caller. Macro parameters can also be names referring to memory locations or even constant names. For example, suppose we want another macro which will add the value in a caller-specified address to a register, then the macro definition would be

```
ADR     MACRO   REG, MEM
        PUSH    PSW
        MOV     A, REG
        LDA     MEM
        ADD     A, REG
        MOV     REG, A
        POP     PSW
        ENDM
```

Whenever the macro call ADR B,X appears in our program the assembler will copy over the six executable instructions from the definition, but replace REG by B and MEM by X where these appear, so that the resulting program segment will add the content of X to register B. Obviously, the symbolic names appearing on the right of the MACRO statement must match those appearing in the macro call, both in number and in type. It would not work if we tried to add two memory locations X and Y using the macro call ADR X,Y, since replacing REG by X would produce incorrect instructions such as MOV A,X.

In addition to permitting the programmer to define his own macros, an assembler normally also has a macro library containing a number of macro definitions for frequently used purposes, especially those related to input/

output or requests to the operating system. Judiciously employed, macros provide the programmer with both flexibility and convenience.

4.4.3 Use of Subroutines

We now discuss the use of subroutines by way of examples. Both macros and subroutines normally have parameters, i.e. data passed from the calling program to the subroutine for processing. There are, however, two important differences. First, in macros the instructions in the macrodefinition are actually inserted into the calling program, so that they are simply part of the program and can access any data available to the program. All that is needed is for the assembler to replace the parameter symbols in the macro definition by the variable names provided in the macro call. Subroutines, in contrast, are physically self-contained and separate, and it cannot be assumed that they have the same data access as the calling program.

It will be necessary to decide on where the calling program will put the data for the subroutine, and where the subroutine will return results, using data storage space that is accessible to both sides. There are a number of such mutually accessible places. Registers, for example, are available to both the calling program and the subroutine, and it is possible for the former to place data or the addresses of data, in some registers for retrieval by the latter. Alternatively, to avoid allocating too many registers, or where the number of data items to be passed over is large, the calling program may construct a list of addresses, and put the starting address of the *list* into a register. To retrieve the data the subroutine would first have to fetch the addresses one by one, and put them into registers, which may then be used to fetch the data. The stack is another place which may be used to store parameter values or addresses.

Agreement is however necessary on the data to be passed and the form of this data. For example, consider the programming example given in the early part of this section to place the value 0 to 99 into a table. Assume that there are several such tables of different sizes and that we wish to write a subroutine for preparing them. The subroutine requires two parameters; the address of the table and its size. These may be passed in the registers B/C and D respectively, and the subroutine would take the following form:

```
SETTBL:   SUB    A
LOOP:     STAX   B
          INX    B
          INR    A
          CMP    D      ; COMPARE A WITH D
          JC     LOOP
          RET
```

the only difference from the original form being the comparison of A with D rather than with a constant, the addition of the subroutine return at the end, and the definition of subroutine name SETTBL so that the calling program may use the subroutine by the following instructions.

```
PUSH    PSW
LXI     B, TABLE
MVI     D, 100
CALL    SETTBL
POP     PSW
```

Here the PUSH and POP save and restore the original value of register A, because it is altered by the subroutine. LXI and MVI place the address TABLE and the size into where the subroutine expects them, and the CALL instruction causes a jump to SETTBL, with the return address (address of POP, the instruction below) saved on the stack, which will later be used by SETTBL to return to the calling program.

However, we now present an alternative way of achieving the same result. Suppose the calling program has the instructions

```
CALL    SETTBL
DW      TABLE
DB   .  100
          .
          .
```

(MORE INSTRUCTIONS)

In other words, instead of putting the address and length into registers, the program locates them below the subroutine jump instruction, since DW causes assembler to allocate 16 bits and to put there a value appearing on the right, which happens to be an address. Thus, the address of TABLE will appear below CALL, DB then causes the assembler to allocate a byte, and 100 is placed there. But how does the subroutine get these numbers? We recall that the execution of a subroutine jump causes the address of the location below to be placed on the stack because this is (normally) the return address. But now the "return address' contains instead the first parameter, namely the address of TABLE, and since the return address is put on the stack by the CALL instruction, the subroutine can obtain the address from the stack, and use it to get the parameters. Thus the following instructions are used to implement the subroutine

```
SETTBL: POP  H   ; RETRIEVE RETURN ADDRESS
        MOV  B,M ; USE ADDRESS IN H/L TO GET FIRST BYTE
        INX  H   ; TABLE ADDRESS, THEN INCREASE
        MOV  C,M ; USE ADDRESS IN H/L TO GET SECOND BYTE
```

```
          INX   H    ; INCREASE AGAIN
          MOV  D,M  ; THEN FETCH LENGTH OF TABLE
          INX   H    ; AFTER THIS ADDRESS IN H/L POINTS TO
                     ; LOCATION BELOW LENGTH, THE PLACE
          PUSH H     ; TO RETURN TO. SO PUT IT ON STACK
          SUB   A    ; THE REMAINDER OF SUBROUTINE IS
LOOP:     STAX B     ; SAME AS BEFORE
          etc.
          RET        ; THIS WILL PICK UP INCREASED ADDRESS
                     ; TO RETURN TO "MORE INSTRUCTIONS" IN
                     ; CALLING PROGRAM
```

We see that many alternative ways are possible, including some which are somewhat obscure at first sight. Note that the above subroutine alters all the registers, and the calling program has to PUSH four times and POP four times both before and after the subroutine call to ensure that no values are lost.

4.4.4 Final Example

We now show a program segment larger than those we have seen so far, both to increase the reader's familiarity with Intel 8080 assembly programs, and to show some of the machine's limitations. The problem to be solved by the program is quite a simple one. Given two vectors V1 and V2, add every second element of V1 to every third element of V2 and put the results into another vector V3. In Intel 8080 assembly code we may write

```
;DATA DEFINITIONS
V1:     DS      200
V2:     DS      300
V3:     DS      100
          .
          .
          .
        ETC.           ; OTHER DATA DEFINITIONS
          .
          .
          .
        etc.           ; INSTRUCTIONS TO READ IN VALUES
        LXI B,V1       ; SET ADDRESS OF VECTOR
        LXI H,V2       ; SECOND VECTOR
        LXI B,V3       ; THIRD VECTOR
        SUB A,A        ; INIT LOOP COUNT. STOP WHEN 100
                       ; REACHED
```

```
LOOP    PUSH PSW    ; SAVE VALUE OF A BECAUSE WE NEED IT
                    ; FOR OTHER PURPOSES
        LDAX\B      ; GET VALUE OF V1 INTO A
        ADD  M      ; ADD NUMBER POINTED TO BY H/L, V2
        STAX D      ; STORE INTO V3
        INX  D      ; INCREASE V3 INDEX BY 1
        INX  B
        INX  B      ; INCREASE V1 INDEX BY 2
        INX  H
        INX  H
        INX  H      ; INCREASE V2 INDEX BY 3
        POP  PSW    ; RETRIEVE VALUE OF A
        INR  A      ; INCREASE LOOP COUNT BY 1
        CPI  100    ; COMPARE WITH LIMIT
        JC   LOOP   ; IF A IS LESS THAN 100 GO BACK TO
          .         ; PROCESS NEXT ELEMENT
          .
          .
        ETC.
```

There are several comments to be made concerning this program.

(a) Since three address pointers are needed for the three vectors, only register A is available for other purposes. But as arithmetic can only be performed on the A register it is necessary to save its value regularly while it is being used for arithmetic purposes and to retrieve the value later. In this program we use the stack although it would also have been possible to define a memory location, say COUNT, and to save A there using STA COUNT and retrieve it with LDA COUNT, instead of PUSH and POP. Note that we would have difficulty in doing this for a macro, since if we have a data definition inside a macro definition the assembler would simply insert the data definition into the calling program along with executable instructions, and a data memory location would exist in the program requiring the programmer to add instructions to jump over the data to the next instruction. For example, ADR of Section 4.4.2 may be re-written as

```
ADR             MACRO    REG
                STA      SAVEA
                MOV      A,REG
                ADD      M
                MOV      REG,A
                LDA      SAVEA
                JMP      SKIP
SAVEA:          DB       0
SKIP:           NOP
                ENDM
```

(b) We have to use H/L to point to V2 because the instruction set only permits adding either a register or the memory location pointed to by H/L into A.

(c) Adding 2 to B/C is carried out by executing INX B twice and adding 3 by having three INX instructions. If vectors are being processed (where, for example, we need to select one element out of every ten), then the programming becomes more complex. The instruction set only permits the addition of constants to register A and to add ten to register B/C we would need to write

```
PUSH   H    ; SAVE PRESENT VALUE OF H/L ON STACK
MVI    H,O  ; PUT ZERO INTO H
MVI    L,10 ; PUT TEN INTO L. NOW H/L HAS 10
DAD    B    ; ADD B/C TO 10 IN H/L
MOV    B,H  ; MOVE FIRST BYTE OF SUM INTO B
MOV    C,L  ; MOVE SECOND BYTE OF SUM INTO C
POP    H    ; RECOVER FORMER VALUE OF H/L
```

It would be slightly easier to add ten to register D/E since we could replace the two MOV instructions by an XCHG instruction which exchanges the contents of register H/L with that of D/E, but the programming is still fairly complex.

(d) We have made the very naive assumption that the elements of all three vectors each consist of one byte. Single byte numbers would be extremely unlikely since their range, -128 to $+127$, is too small to be of much use. Extending the program to 2-byte or 4-byte numbers would be difficult because multiple-byte addition has to be built up using single-byte additions, e.g. for double-byte numbers the program would be modified as follows:

```
LOOP   PUSH   PSW
       LDAX   B
       ADD    M      ; ADD LESS SIGNIFICANT HALF
       STAX   D
       INX    B      ; INCREMENT REGISTERS TO
       INX    H      ; POINT TO SECOND HALVES
       INX    D      ; OF VECTOR ELEMENTS
       LDAX   B      ; FETCH SECOND HALF FROM V1
       ADC    M      ; ADD V2 WITH CARRY FROM FIRST HALF
       STAX   D      ; STORE INTO V3
       INX    D      ; INCREASE V3 INDEX BY 2−1=1
       INX    B      ; INCREASE V1 INDEX BY 2×2−1=3
       INX    B
       INX    B
       etc           ; INCREASE V2 INDEX BY 2×3−1=5
```

Note that if elements are two bytes in length, then taking one value out of every three means 2 bytes out of every 6, which is why the index increases have to be changed. As each index has already been increased once, this has to be taken off the required increase. Hence, at the end of each loop, register B/C ought to go up by 3 and H/L by 5.

(e) We have also done a number of things which are not the best from the point of view of programming methodology. Instead of putting 100 into the CPI instruction it would have been better to define a constant, say LIM EQU 100, and use the name in the program to facilitate change of vector sizes. We have not considered the possibility that vector sizes are non-constant,—we might write a subroutine which is called several times, each to process different vectors of varying size. It is also possible for the vector size to exceed 256, so that it cannot be stored in one byte, and a simple CPI instruction no longer suffices. Further complication may arise if the index increments are to be variable, i.e. add one element out of every N from V1 to one out of N from V2 and put the sum into one out of L elements in V3. Each will require substantial code modifications to the program.

It is our hope that we have succeeded in giving the reader a glimpse into the tedious, error prone, but often fascinating process of microprocessor assembly programming. A newcomer to the subject is likely to find it difficult initially to consider the large number of instructions and pseudo-operations and to know when to use them or to understand the exact format and restrictions on their applicability. However, once familiarity is gained with one processor it is not too hard to follow the assembly languages of other micro-processors. Learning to program efficiently in assembly code is a little more difficult however. Several publications covering this subject in more detail are suggested at the end of the chapter.

It should be stressed that assembly code should be reserved for those problems in which consideration of processor or I/O efficiency makes the use of high-level languages undesirable. Not only are high-level language pro-grams easier to write, understand and modify, leading to greater programmer productivity through reduced tedium and frustration, they are more trans-portable to other processors, so that rewriting of software is reduced should hardware changes become necessary.

4.5 THE SOFTWARE DEVELOPMENT PROCESS

In this section we outline briefly the process of software development and indicate the facilities required. Details of software development aids will be discussed in Chapter 6.

As stated, the programmer writes his programs in either the assembly language for the microprocessor or some high-level language, using a notation and a system of naming conventions which make the program more comprehensible and therefore easier to write and correct. However, such source programs cannot be executed directly by the processor. Instead, they must be translated or *compiled* into machine instructions. The translation process is highly complex and takes a computer to perform rapidly and correctly, and the program that does such a job is called a *compiler*. Since a compiler is designed to accept programs written in a particular language and produce machine programs for a particular processor, any user who wants to run a variety of languages on a variety of processors will need a huge range of compilers. This is why some form of standardization is highly desirable, both to reduce the range of software facilities required and to make programs transportable from system to system.

Since a compiler is a computer program, it requires a host computer system. Further, the system must satisfy certain minimum requirements for software development to be possible. A compiler is usually a fairly large piece of code, and takes much memory, both to store the code itself (or a substantial portion of it, with other parts brought into memory as needs arise), and to store the information it needs to perform the translation. There has to be a reasonably fast input device on which the source program may be read in for translation, and some output device on which the machine program is prepared by the compiler, in a form suitable for loading later when processor and memory are available to execute the program. There must be facilities by which the programmer may prepare his program for input to the compiler, and to make changes. As there may be a large number of programs which are either in production or in various stages of development, it is desirable that they should be stored on the system in some readily accessible form, and be available for inspection, change or execution as need arises. It is also desirable to have facilities for test executions of a program under development, with the programmer interactively changing parts of the code or its execution environment and observing the resulting effects.

Because microprocessor systems like those used in signal processing applications, rarely have the complete range of hardware and software facilities required by the software development process, frequently the software has to be developed on a separate, much larger computer, and the machine code transferred across after development to be executed in the microprocessor itself. A range of such facilities, including cross-compilers, emulators, editors and file managements systems, will be discussed in Chapter 6.

However, the increased power and I/O handling capabilities of microprocessors, together with decreasing cost of microprocessor I/O and software packages, have made it increasingly economical to provide software development facilities on microsystems themselves. Today many stand-alone microcomputer models are available which have a floppy-disc based operating system, with compilers for a small range of languages such as Assembler, Basic or Pascal, and simple editing, file management and debugging facilities. Two widely adopted operating systems are CP/M (Control Program for Micro processors) and UCSD Pascal System. With both systems, a user may input his program on the system keyboard, to be stored on disc as a program file. He can inspect the program on the screen or printer, modify it using the editor, make copies or dispose of unwanted programs using standard system requests, compile his programs, or initiate execution using standard console commands. Whereas CP/M is a fairly simple but unsophisticated system, UCSD Pascal is a more polished product, but is limited to supporting Pascal. However, many microprocessor systems may switch from one to the other through a simple reload process, and perhaps offer other manufacturer-specific systems as well.

4.6 DIRECTIONS FOR MICROPROCESSOR ARCHITECTURE DEVELOPMENT

Despite the effectiveness of microprocessors for solving relatively simple data processing problems, as illustrated by the programming examples earlier, generally speaking microprocessor programming can be a tedious and often frustrating task. Part of the problem lies in the low range and power of programming tools, on which further discussion will be given in the next two chapters, and part of it is due to the limitations of microprocessor architectures and instruction sets. In this section we comment on these problems and explore directions of microprocessor architecture development needed to overcome them.

The final programming example in Section 4.4 illustrates two main difficulties; the small number of registers available, and limited freedom for data manipulation operations on data stored in places other than the accumulator. These two problems are due to the same cause, namely the small number of instruction formats available with 8-bit instructions. Out of the 256 available instruction formats, 63 are required for the purpose of moving data between the seven registers and the memory. With a larger number of registers, even fewer instructions would be available for other purposes. This explains why the Intel 8080/85 arithmetic instructions always operate through the accumulator. If arithmetic needs to be carried out in the context of all seven registers then the arithmetic instructions will have to contain some bits to indicate which register is to be used and hence these bits will cease to be available for other purposes. The Intel 8080/85 designers could not follow this strategy, because they needed to provide a large number of other instructions to ensure a viable general-purpose processor, so that reduced freedom in arithmetic operations was the result. Similar restrictions were imposed on the store direct instructions, again because of the shortage of instruction formats.

The same problem is present in the Motorola 6800 processor but in a different form. As mentioned in Section 4.1, this processor has special rather than general purpose registers (two 8-bit data registers and one 16-bit address register). Since its design ensures that data-manipulation type instructions only refer to data registers and address-type operations only operate on the index register, the programmer faces similar limitations. Further, the 6800 has the ability to perform arithmetic on both data registers, and the instruction set provides for a number of arithmetic operations combining data in a register with that in a memory location. On the other hand, the Intel design permits the use of its six registers as three 16-bit address registers (B/C, D/E and H/L), and may store up to 6 frequently used numbers in the registers B to L for complex combination with the contents of register A.

We see from this that given the ultimate instruction format restriction, gains in programmer freedom in some areas have to be paid for by loss of freedom elsewhere. In fact, the only workable solution to the problem is an increase in the instruction length. In the case of Z80, the increase is implemented in a selective fashion. All the standard Intel 8080 instructions functional in the Z80 design retain the 8-bit format, but a large number of additional two-byte instructions are added. With the Z8000 and other modern microprocessors, a full 16-bit instruction format is implemented, to provide for the specification of 65536 different operations (or the same operations using different operands).

However, a careful study of the instruction sets for this new range of 16-bit microprocessors will show that the instruction format limitation continues to present a problem. For, with a more powerful processor, the designers are obliged to provide a greater number of registers and an expanded instruction repertoire, so that the improved result is essentially a compromise. Taking the Zilog Z8000 as an example, this has 16 general purpose registers, and a two-operand instruction which requires four bits each for source and destination fields. Of the remaining 8 bits, 6 are needed for the opcode, since a long opcode is required in order to provide a variety of instructions, in particular byte, double-byte and 4-byte operations of various kinds. This leaves only two bits for addressing mode specification. At the same time, to make the machine truly versatile, the designers provide no less than seven addressing modes: register, immediate, indirect, absolute (direct), indexed, base and indexed base. (The last two modes are somewhat similar to indexed mode, but are related to the needs of the segmented memory addressing of the Z8000). Since 2 bits are insufficient to specify seven modes, a complicated set of rules are imposed on the availability of each addressing mode with individual operations or registers. In consequence, the same addressing mode value may mean several different things depending on what the instruction is and which register it uses, and the programming complexity is accordingly increased.

We can also identify a second, related, problem. The modern microprocessors are required to support larger system configurations, particularly in terms of memory size. Thus, both the Zilog Z8000 and the Motorola 68000 support 24-bit addresses, and with 16-bit instructions and access to common datatypes, a 16-bit data bus is required. While an 8-bit data bus is feasible, this greatly reduces system speed as instruction and data fetches become much slower (as in the case of the Intel 8088). We recall, however, that a microprocessor chip can only have a limited number of pins. The Motorola 68000, having separate address and data pins, has a 64 connector IC package, and the large chip size makes the product more costly to produce. The Zilog Z8000 and Intel 8086, on the other hand, have reduced chip sizes, but at the cost of having shared address and data pins. This makes the traffic control circuits both inside and outside the processors more complex, since addresses and data pulses have to be channelled to the same lines at one end and correctly separated at the other end. Further, the arrangement limits the realizable speed of the system since the processor cannot place data and address pulses on the system bus simultaneously, but must adopt a execution cycle structure in which data and addresses are transferred at different steps in a cycle.

In summary, the above discussion indicates that, while it might be considered desirable to increase further the power and versatility of microprocessor instruction sets by increasing the word length beyond 16-bits it is apparent that this results in diminishing returns for the cost of achieving this. In consequence, it is useful to ask whether there may be other approaches, which will improve performance *within* the limitation of short instruction lengths. It is pointed out in [10] that advantage may be taken of the fact that frequently successive instructions in a program refer to the same register or a set of closely associated registers, and one can design 8-bit processors with larger amounts of register storage than are presently available to include a more versatile instruction set, if one based the architectural design on families of registers and context-dependent machine instructions having implicit register references. The same idea may also be applied to larger processors to increase further their capabilities. While the current trend towards longer instructions and addresses, increased virtual and physical memory size, greater I/O capability, and physically bigger processors with more pins, will continue, it should be possible to achieve qualitative improvements resulting from these quantitative increases.

References

1. Laventhal, L. A. (1978). "Introduction to Microprocessors: Software, Hardware, Programming". Prentice-Hall, New York.
2. Mateosian, R. (1980). *Programming the Z8000*. Sybex Corp., Berkeley, California.
3. Zaks, R. (1979). *Programming the Z80*. Sybex Corp., Berkeley, California.
4. *Component Data Catalog* (1981). Intel Corp., Santa Clara, California.
5. *Microprocessor Application Manual* (1975). Motorola Inc. and McGraw-Hill, New York.
6. *MC68000 16-bit Microprocessor User's Manual* (1980). Motorola Inc., Austin, Texas.
7. *Z8000 Technical Manual* (1979). Zilog Inc., Cupertina, California.
8. Peuto, B. L. (1979). Architecture of a New Microprocessor, *Computer* 12(2), 10–21.
9. Stritter, E. and Gunter, T. (1979). A Microprocessor Architecture for a Changing World—The Motorola 68000, *Computer* 12(2), 43–52.
10. Yuen, C. (1981). Extending the Power of Short-wordlength Processors by Means of Context-dependent Machine Instructions, *Computer Architecture News* (October).
11. Zaks, R. (1980). *The CP/M Handbook with MP/M*. Sybex Corp., Berkeley, California.
12. Bowles, K. L. (1980). "Beginner's Guide for the UCSD Pascal System". Byte Books, Peterborough, New Hampshire.
13. Jackson, M. A. (1975). "Principles of Program Design". Academic Press, New York.
14. Yourdon, E. and Constantine, L. (1975). "Structured Design". Yourdon Press, New York.
15. Myers, C. J. (1975). "Reliable Software Through Composite Design". Petrocelli/Charter, New York.
16. Yourdon, E. (1979). "Structured Design: Fundamentals of a Discipline of Computer Program and System Design". Prentice-Hall, New York.
17. Dagless, E. L. and Aspinall, D. (1982). "Introduction to Microcomputers". Pitman, London.
18. Osborne, A. and Kane, G. (1981). "4 and 8 Bit Microprocessor Handbook" (Vol. 1), "16 Bit Microprocessor Handbook" (Vol. 2). Osborne/McGraw Hill, Berkeley.

5 Microprocessor Input/Output Handling

5.1 SIMPLE I/O SYSTEMS

In this section we introduce the basic concepts of microprocessor I/O systems, both hardware and software. As indicated in Chapter 1, the subject involves a large number of complex details varying from machine to machine and from device to device. Much of such detail should be the concern of electronics technicians rather than that of system designers and users. However, by familiarizing himself with the material presented here, the reader will greatly improve his ability to communicate effectively with the technical personnel.

We start with simple I/O systems. By this term, we mean systems in which the I/O devices play a purely passive role, and all the detailed operations involved in an I/O process are controlled fully by the processor, which is dedicated to this task. Thus, there is no concurrent execution of tasks unrelated to the I/O operations.

To enable such I/O operations to occur, a number of capabilities must be provided in the hardware, both in the processor and in the I/O subsystem. The processor design must provide for the execution of instructions which will request specific I/O operations by producing appropriate pulses on the system bus. Some of these pulses indicate to the system that an I/O operation is being requested, other pulses (I/O device address) specify which device is required, and yet others specify the particular operation needed, e.g. READ Status, OUTPUT Control, INPUT Data or OUTPUT Data. In Addition, the I/O subsystem must contain the following three facilities; first is I/O operation recognition, in which the subsystem recognizes that there is an I/O request from the processor rather than a memory or some other operation; second is device selection, in which the I/O subsystem identifies the particular device to be accessed using the address information provided by the processor, to open appropriate information paths between the processor and the device. Finally there must be signal conversion, in which the I/O subsystem generates the control signals needed by the I/O device to carry out the operation specified by the processor and to transfer the data or status information through the subsystem, making the necessary changes in the form of the data/status pulses in cases where the input or output routes have different operational requirements.

As each microprocessor model has its own way of specifying an I/O request, and each different device has different control and status designs, the functional requirements placed on an I/O subsystem varies greatly. Further, there is a wide choice of implementation methods. One can, for example, design a complete I/O subsystem using the basic digital components, particularly in order to incorporate special devices into the system. However, it is usually possible to adapt some of the interfacing modules provided by the manufacturers to meet normal device handling requirements. Another decision the designer has to make is the level of I/O control centralization. In the decentralized (or *unibus*) scheme the I/O request pulses provided by the processor are sent to all I/O devices, and the logic within the system interface performs the I/O request recognition, device selection and signal conversion: Each device must contain sufficient intelligence to decide whether it is being asked to do something, and if so what is to be done and how to do it. The result of the operation is then placed on the system bus for transmission back to the processor. In the centralized (or *system controller*) scheme the I/O control module alone receives the I/O request pulses from the processor, decides what needs to be done, then sends the necessary control pulses to the particular device requested to effect the I/O operation, before transferring the result back to the processor. (In many systems we find some middle ground between the two schemes.)

Besides the hardware provisions, a great deal of software is also needed to utilize I/O devices, since the operation of each device requires particular combinations of status detection, data transfer and control pulse output. The software becomes especially elaborate when the hardware provides more than just simple I/O handling, as we shall see later when discussing interrupts and direct memory access.

We shall again base most of our discussion on the Intel 8080, despite the fact that it is superseded by Intel 8085. This is because the latter incorporates a major part of the I/O interfacing logic into the processor design itself, and hence does not provide the opportunity of using it to discuss how the logic works by referring to individual components. We shall, however, make brief mentions of the I/O design for other microprocessors where such discussions are particularly illustrative. Particular attention will be paid to the differences between Intel 8080 and 8085 I/O handling.

5.1.1 Basic I/O Hardware

We mentioned in the previous chapter the two Intel 8080 I/O instructions IN and OUT, 11011011 and 11010011 respectively. Each will be followed by the I/O port number identifying the device that the instruction is meant to access. During the execution of IN or OUT the following sequence of events will occur

(a) In the first step of execution the processor places an address on the address lines, and a functional request (called in Intel terminology *status information*) on the data lines D_0 to D_7. An IN instruction causes D_6 to be raised, while an OUT instruction raises D_4. The I/O subsystem, upon receipt of either pulse, will recognize an I/O request. If instead of D_4 or D_6 the processor raises pulses on other data lines then the interface will recognize non-I/O operations, e.g. D_5 indicates an instruction fetch, D_7 indicates a memory READ for data, D_2 a stack operation (either READ or WRITE), D_1 a memory write, while D_3 denotes a program HALT. (The purpose of D_0, interrupt acknowledge flag, will be explained in the next section. We have also simplified the details of the status information presentation, so that the above can only be considered as an approximate representation).

(b) In case of output, the processor then places the 8-bit data for the selected device on the data lines, while at the same time activating the WR line, which is one of the control lines provided in the 8080 chip design (see Fig. 4.2, pin 18). When the I/O subsystem receives the WR pulse, this indicates that output data have become available, and the values on the data lines are switched to the device whose address was provided during step (a). In case of input, the processor merely raises the DBIN pulse (pin 17), to indicate that it is ready to receive data. Upon receipt of this, the I/O subsystem switches the 8-bit data from the device to the processor data lines. Again the above description is an oversimplified picture but is sufficient for our needs.

It should be noted that not only does the I/O subsystem have to perform the necessary I/O request recognition, device selection and signal conversion in response to a processor request, it has also to do this within specific time limits. Normally the Intel 8080 operates at the clock rate of 2 MHZ, or at 500 ns per machine cycle. Each of steps (a) and (b) lasts about 200 ns, and the I/O subsystem has to respond to the processor signals within these time limits. In fact each step may contain smaller sub-steps, within which certain definite actions are required from the I/O system. The necessary synchronization is controlled by a system clock module (usually the Intel 8224 chip), which produces two clock pulses per machine cycle, for input into the processor pins ϕ_1 (pin 22) and ϕ_2(17). Depending on which pulses are present, the processor is made to enter different steps of instruction execution, synchronized with the I/O and memory subsystems.

It is also relevant to note that instruction fetches and executions involving memory accesses follow similar steps to I/O operations. In step (a), a different functional request will appear on the data lines, and a memory address, rather than a device number, will be placed on the address lines. The request activates the memory subsystem instead of the I/O subsystem and, during step (b), the WR or DBIN pulse cause the transfer of data between a memory location and the processor.

We now turn to the capability required of the I/O system. It is required to activate a device which meets the following conditions.

(1) For an input device $D_6 = 1$ during step (a). For an output device $D_4 = 1$. No additional circuits are required for the recognition of an input or output request from an Intel 8080 processor. However, in alternative processor systems, I/O requests may take other forms and recognition circuits will be needed.

(2) The device address appears on the address lines. The requirement of device selection may be met either by a decoder, which will send a pulse to one out of 2^n alternative devices depending on the value appearing on n input lines. However, it is also possible to perform device selection by building address-recognition circuits into each device interface, and connecting the processor address lines to all devices, such that the device whose number appears on the lines will recognize it and respond.

(3) For an input device, activation occurs when DBIN $= 1$; for an output device it occurs when WR $= 1$.

Figure 5.1 shows the logical structure of a I/O system controller capable of handling 8 input lines and 8 output lines. It contains a 3-to-8 decoder, which accepts address lines A_2 to A_0 as input and selects one out of 8 alternative control lines, each of which is used to control one input port and one output port. An input control pulse is sent if D_6 was 1 during step (a), followed by DBIN $= 1$, which opens the AND gate connected to the input port. Note that $D_6 = 1$ occurs before DBIN goes to 1, by which time the data lines have taken on new values. Hence, the value of D_6 during step (a) is saved in a flip-flop for use during step (b). An output control pulse is sent to the device whose address is specified if $D_4 = 1$ during step (a) and WR $= 1$ during step (b). Again, D_4 has to be saved in a flip-flop. Note that, since only one out of the 8 outputs of the decoder can be 1, and D_4 and D_6 cannot both be 1, only one out of the 16 AND gates would be open at any time, so that only one device would be selected.

This describes the *logical* design. In actual implementation it is more usual to employ a number of manufacturer-supplied modules to achieve the same

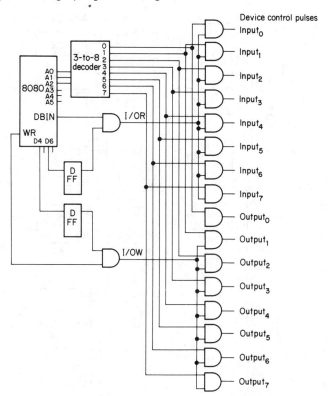

Fig. 5.1. Control systems for 16 I/O ports.

logical functions in a manner more convenient to the user. Figure 5.2 shows such an I/O control system, implemented using an Intel 8228 *system controller module*, a 2-to-4 decoder, and a number of Intel 8212 *I/O ports*. To explain the operation it is necessary to describe the internal constructions of the two Intel modules illustrated in Figs 5.3 and 5.4. There are three main components in the 8228 module: A *bidirectional data bus* which connects the data lines of the 8080 processor with those of the I/O devices, to provide two-way traffic and performing the necessary signal conversion and stabilization; a *status latch* which stores the values of D_0–D_7 produced by the processor during step (a) for use during step (b); and a number of gates which produce I/O and memory control pulses from the stored D-values and additional control pulses from the processor during step (b), such as WR and DBIN. The 8212 I/O port, illustrated in Fig. 5.4, is basically a dual-controlled gateway for 8 data lines, which may be set in one direction for input and in the other for output depending on the value of the mode (MD) input. The

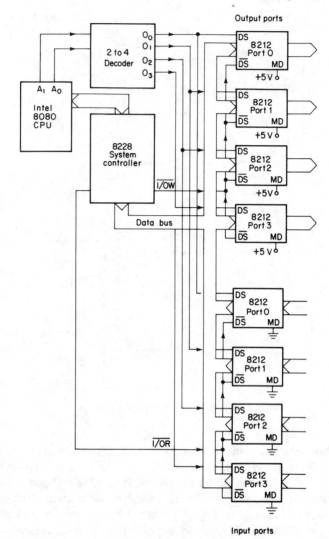

Fig. 5.2. Intel 8228 Controller based system. (With acknowledgements to Intel Corp.)

opening or closing of the I/O port is controlled by two input bits; one from the decoder for device selection, and the other from the system controller for I/O recognition and step (b) timing. The overall system provides the same logical relations as those of Figure 5.1, but with enhanced capabilities to assist memory interfacing and data bus driving.

In the preceeding paragraphs we have considered I/O devices and I/O

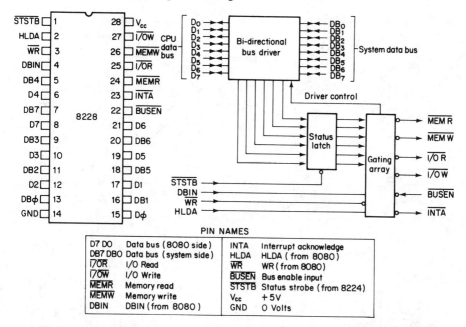

Fig. 5.3. Internal logic of 8228 controller. (With acknowledgements to Intel Corp.)

ports without discussing their separate meanings. It is usually necessary to connect an I/O device through several ports because we need separate paths for data, status and control, and some devices operating with longer word-lengths than that of the processor may need several datapaths. A device with 16-bit data, up to 8 status lines and control lines would be connected to one input port for status, one input port for control, and two input or output ports for data depending on whether it is an input or output device. Thus, although the Intel 8080 is able to control 256 input lines and the same number of output lines, the total number of *devices* it can handle is much less than 512. To use an I/O device, the Intel 8080 program has first to execute an IN instruction from the status input port for the device and to analyse the information to see whether the desired operation may be performed, and if so, to execute the OUT instruction to the control port for the device, sending to it the necessary control pulses, followed by more IN or OUT instructions to transfer the data. We shall consider details of I/O programs in the next subsection.

This concludes our discussion on basic I/O hardware for the Intel 8080. While actual systems can be highly complex, the foregoing discussion has illustrated the important concepts involved. We shall later see that the ideas

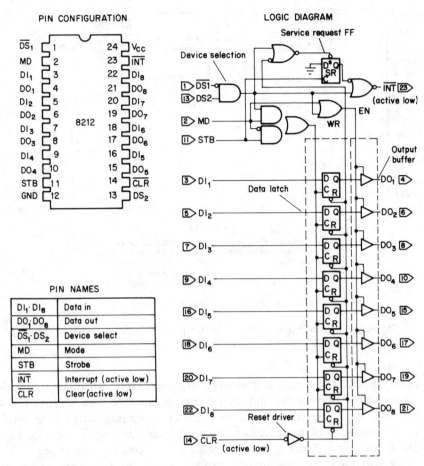

Fig. 5.4. Intel 8212 I/O port. (With acknowledgements to Intel Corp.)

carry over to other Intel microprocessors (section 5.1.3), as well as to processors in general which have special I/O instructions rather than memory-mapped I/O.

It is, however, useful to discuss briefly the I/O configuration of the Motorola 6800, which does not have separate I/O instructions so that I/O transfers are performed quite differently using memory READ/WRITE instructions. I/O request recognition is thus dependent on the value of the address; for example, the system designer may designate the addresses above 2^{15} as being reserved for I/O devices, so that if bit 15 of the address lines is 1 then a I/O transfer is being requested. However, this limits severely the total amount of memory one can connect to the system, since only the addresses 0 to $2^{15} - 1$ are available for memory locations. A much smaller range of address values is assigned to I/O devices and I/O request recognition is more difficult to achieve. We find that Motorola 6800 systems tend to adopt a more decentralized form of I/O control. The manufacturer supplies a number of device interface modules which incorporate such functions as I/O request recognition, device selection and signal conversion into the device interface. A commonly employed module is the 6820 peripheral interface adaptor (PIA), which has a high degree of "internal intelligence". By connecting a PIA to selected address lines of the processor, the adaptor will respond to only certain values of address coming from the processor. Each adaptor has two sets of registers, for handling two devices. By placing the appropriate values into some of these registers, the processor can make the adaptor perform either input or output operations in some specified manner or to return operational information. The processor can also control whether the devices on the adaptor are permitted to cause an interrupt (to be discussed in the next section).

It is clear from the above that I/O handling systems can be designed in a number of ways, each having its own advantages and disadvantages. With standard I/O devices each manufacturer recommends a given technique for their implementation, and if this is followed few problems occur. For less standard devices the Intel 8080 type system normally requires the development of special interface circuits, whereas the Motorola 6800, because of the greater flexibility and intelligence of its PIA, can usually handle the special requirements of the device through suitable programming. Where this fails to provide the necessary control mechanism, so that the design of a direct path to the processor using memory mapped systems becomes necessary, then this becomes more difficult.

5.1.2 Simple I/O Programming

In this section we consider the I/O programming of a few slow devices which produce or accept data one byte at a time (i.e. not block-oriented devices like disc transports). First consider a slow printer. Such a device requires two I/O ports, an output port connected to the data lines of the printer, such that the processor will cause a character to be printed if the appropriate 8-bits are available on that port, and an input port connected to the status lines of the device. There is no need for control lines, since sending a character to the printer automatically causes its printing, and no other control operations are available for the processor to request. One status bit at least is necessary however, since it takes some time (typically 1/30 sec.) for the printer to do its job, during which period the processor must refrain from sending out the next character to avoid interfering with the printing of the previous one. Let us assume that the BUSY/FREE status of the device is indicated by the zero bit of the status lines terminating at input port No. 1, whereas the data lines connect to output port No. 1. A complete output process then consists of the following steps

A. Read device status and test BUSY/FREE bit;
B. If BUSY repeat from step A, otherwise proceed to step C;
C. Output byte to device data buffer.

In Intel 8080 assembly language, the following code segment implements this process repeatedly, to input a block of characters, whose starting address is in registers H/L and length is in register B.

```
TEST IN    1     ; INPUT DEVICE STATUS FROM PORT 1
                   INTO ACCUMULATOR
     ANI   1     ; TEST BIT 0 OF DEVICE STATUS BY
                   ANDING ACCUMULATOR WITH CONSTANT 1
     JNZ   TEST  ; IF RESULT NOT ZERO THEN BIT 0 WAS
                   1 HENCE DEVICE BUSY. REPEAT TEST
     MOV   A,M   ; MOVE BYTE FROM MEMORY LOCATION
                   POINTED TO BY H/L TO ACC.
     INX   H     ; INCREASE ADDRESS BY 1
     OUT   1     ; OUTPUT BYTE IN ACC. TO PORT 1
     DCR   B     ; DECREASING CHARACTER COUNT
     JNZ   TEST  ; IF NOT ZERO, GO PRINT NEXT BYTE
```

It might be asked why it is necessary to keep testing the status until the device becomes free. As indicated earlier in Chapter 1, the timing of the I/O devices are difficult to predict and control. While a printing operation typically takes 1/30 of a second, the actual amount varies. The processor can control when the device enters BUSY state (by giving it a character to print) but not when it exits from that state. Some repeated checking cannot be avoided if we are not to use more elaborate mechanisms than the basic hardware so far described. However, it is often possible to reduce the amount of processor executions wasted in repeated status testings by inserting some brief, non-I/O processing instructions, e.g. instead of a JNZ (jump on no zero) instruction to TEST one puts in a subroutine call to some code which does some work unrelated to the printing and timed to finish in about 1/30 sec. By the time control returns to TEST the printer would either have finished printing the received character, or finish shortly. Unfortunately it is not always feasible to find such convenient processing tasks, so this technique is not often applicable.

Now let us consider a simple input device; a keyboard for typing in characters by hand. This is somewhat more complex to program than a printer, which may be activated simply by sending to it a character so that no special control operation need to be performed by the program. A keyboard, on the other hand, is controlled from two sources; the processor might request a character input, but this will actually take place only if the person at the keyboard pushes a button. A keyboard has to be interfaced to three I/O ports; an output port to receive control pulses from the processor, an input port to return the status information and another to return data. The status is also more complicated. An input request from the processor would put the device into the BUSY state, to notify the processor not to attempt second input. A second status bit is also needed to tell the processor whether a character has been input by the human operator. This is provided for by adding a READY/NOT READY status bit, which is turned on as soon as a button is pushed on a keyboard and a character is placed in the data buffer. When the processor, in response to the new status, reads the byte in from the device, both BUSY and READY bits are automatically turned off, to notify the processor that there is nothing left for it to fetch, and that a new input may be initiated.

We present below a code segment which is designed to input a complete line of text from the keyboard. The input processor terminates either when 80 characters have been received, or when the Carriage Return symbol is received (00001101, or decimal 13). Note that if neither occurs then the processor will never exit from this code segment.

```
        MVI   C,13    ; PUT CR INTO REGISTER C
        MVI   B,80    ; MAXIMUM LENGTH OF TEXT
TEST    IN    2       ; INPUT DEVICE STATUS FROM PORT 2
        ANI   1       ; TEST BIT 0 FOR BUSY
        JNZ   TEST    ; IF BUSY REPEAT TEST
        MVI   A,1     ; PUT 1 INTO BIT ZERO OF ACC
        OUT   2       ; SEND THIS TO CONTROL PORT OF DEVICE
                      ; TO ACTIVATE LINE 0
WAIT    IN    2       ; READ DEVICE STATUS
        ANI   2       ; NOW TEST BIT 1 FOR READY
        JZ    WAIT    ; IF ZERO NOT READY SO REPEAT
        IN    3       ; INPUT BYTE FROM PORT 3
        MOV   M,A     ; PUT INTO MEMORY
        INX   H       ; INCREASE ADDRESS
        CMP   C       ; COMPARE WITH CR IN C
        JZ    EXIT    ; IF EQUAL EXIT FROM LOOP
        DCR   B       ; DECREASE COUNT
        JNZ   TEST    ; IF NOT YET ZERO TRY NEW INPUT
EXIT    ...
```

Although we used Intel 8080 assembly language for illustration, the I/O operations on other processors are largely similar. On Motorola 6800, for example, the above program structures would work with appropriate changes of detail. I/O ports would become memory location addresses, and MOV instructions would replace by Motorola equivalents. Basically the idea of status testing alternating with I/O transfers is essential to all simple I/O systems.

5.1.3 Intel 8085 and Other Manufacturers' I/O Systems

The Intel 8085 design attempts to take advantage of the increased chip complexity which became possible after the 8080 development. The processor contains an internal clock generator, I/O and memory control signal generators (somewhat similar to those of the 8228 system controller), and additional interrupt handling circuits. A microprocessor system based on an 8085 tends to contain fewer components than one based on an 8080 because of the increased capabilities of the processor, which make it unnecessary to include an 8228 chip and reduce the complexity of interrupt logic. However, because the the 8085 has a larger number of control signal lines, its 40 connector pins are assigned very different functions from the 8080 assignment. Whereas the 8080 has separate pins for address and data (16 for the former and 8 for the latter), with the 8085 the data pins are combined with 8 of the address pins, in order to free eight pins for other purposes. Thus, whereas in the execution of an 8080 instruction the processor maintains the value of the memory or I/O address on the address lines for the whole period, with an 8085 half of the address bits are maintained only during the early part of the execution; the processor will then remove the eight address bits, and use the eight lines to pass data bits instead. So before the address bits are removed, the I/O and memory systems must save them in an eight-bit register; otherwise they would not be able to recall which memory location or I/O device the processor wanted to access. Whereas standard I/O ports and memory chips do not have such a capability, Intel supplies two special modules, the 8156 and 8355, that may be directly connected to an 8085. The former module contains 256 bytes of RAM together with three I/O ports, and the latter has 2048 bytes of ROM and two I/O ports. Both have been designed in such a way that the address/data lines and control lines of the 8085 may connect to corresponding pins of the 8156 and 8355, without requiring intervening logic.

Provision, however, has also been made to interface the 8085 with standard memory or I/O chips. This is illustrated in Fig. 5.5. Here we have the 8 combined address/data lines (AD lines) of an 8085 connected directly to the 8 data lines of a memory module and an I/O module, and also connected

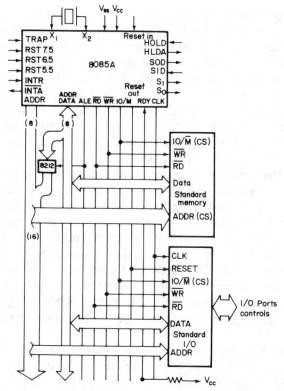

Fig. 5.5. Intel 8085 Processor System. (With acknowledgements to Intel Corp.)

indirectly to 8 address lines of these modules via an 8212 port, which, as we recall from earlier discussion, contains 8 storage elements. These are loaded at the early stage of an execution by the ALE (address latch enable) pulse coming from the 8085, which initiates the signal at the same time as it places address bits on the address lines. The 8212 then holds the address bits for the remaining period of the execution, making them available to the memory and the I/O systems. In contrast, with an 8080 system the address and data lines of the processor would connect directly to the corresponding lines of the memory and I/O modules, but the control signals will come indirectly, via an 8228 controller.

I/O Interfacing for the later Intel microprocessors, 8086 and 8088, is rather more complex. Both have 16-bit registers and instruction formats, and the instruction sets of the two machines are very similar. However, the 8088 has an 8-bit data bus, while the 8086 uses 16 bits for data. Each processor has

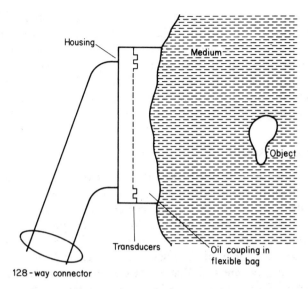

Housing
Medium
Object
Transducers
Oil coupling in flexible bag
128-way connector

Fig. 9.3. An array of transducers used for body scanning.

ter/receiver pair previously considered. Such an array is shown in Fig. 9.3. This consists of 128 tiny transducers mounted on a backing block. Close coupling to the body contours is ensured by the use of a fluid-filled bag. The fluid (usually oil) is chosen to maximize energy transfer into the body under examination.

Early electronic scanning technique used transistor switches to couple each of the transducers to the transmitter/receiver circuits. (The same elements are used in both transmitter and receiver). This provides an electronic analogy to the B-scan system, whereby the beam is moved rapidly across the area of investigation at a rate determined now entirely by the time taken for sound to travel to the most distant point of investigation and back. Since sound travels at a rate of about $1 \cdot 5$ mm/μs in body tissue, a scan covering a depth of 20 cm would require 266 μs per transducer element, resulting in a complete 128-line scan in 35 ms, which is adequate for a refreshed television-type display. Present use of such techniques is such that reasonably high-definition real-time scanning for obstetrics and cardiology may now be realized.

A difficulty from the medical diagnostic point of view, is that whereas the techniques so far described produce an image which represents a "slice" through the body in a place which includes the transducer itself, what is actually required is an image represented by a plane parallel to that of the transducers and at a selected depth. To achieve this it is necessary to employ two separate arrays, one for transmission and one for reception of the acoustic signals. In the system we shall be describing the transmission method involves scanning and focusing of the transmitted signal by means of a *phased array* and the receiving method requires a *holographic reconstruction* of the reflected signals. Only a broad outline of the physical system is given here. Readers who require further details are directed to the reference given at the end of this Chapter [2].

9.1.2 Phased Arrays

The term phased array is used to describe the use of an array of transducers to produce a beam of sound whose direction of propagation can be controlled electronically. Fig. 9.4 illustrates the principle of constructive interference between wavefronts propagating from an array of transducers. At a large distance from the array, the radius of curvature of the wavefronts is so large that waves appear planar. If the phases of the signals driving the individual elements of the array are such that all the elements are driven in phase, then the direction of propagation will be normal to the array. However, if a progressive phase shift is imposed upon the driving signals such that each differs from its neighbour by an amount, ϕ degrees, then the wavefronts will add along a line at this angle to the X axis. It can be shown [1] that the angle

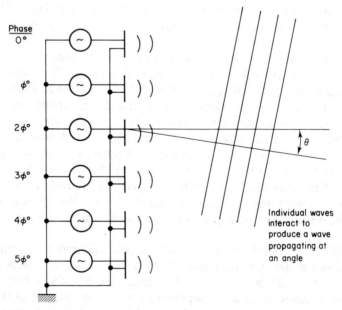

Fig. 9.4. Constructive interference between wavefronts.

of divergence, θ, from the axis is proportional to the phase difference ϕ between adjacent elements.

In order to increase the "resolution" of a phased array it is necessary to focus the beam produced. This can be achieved by allowing the inter-element phase shifts to be adjusted in such a way as to cause the signals from the array to converge at the desired target. The beam-forming calculations required for each of the 128 different array elements over the period of scanning search represents an important task for a microprocessor incorporated in an imaging system, as we shall see later.

In Fig. 9.5 we see the effects of simultaneously focusing and deflecting the beam of sound. Here the phase shifts necessary to focus the beam at a given depth D from the array are superimposed upon those phase shifts needed to deflect the beam through a given angle θ. It can be shown [1] that the electrical phase angle required between a given element and its neighbour is given by

$$\beta = k(r_n^2 + d^2 + 2r_n^2 d \sin \theta)^{1/2} \tag{9.1}$$

where β = electrical phase angle; r_n = distance from the nth transducer to the focal point; d = inter-element spacing; θ = required deflection angle. This assumes that the focal point is in the far field of the array (i.e. at a considerable distance from the array). For close objects (in the near field) the focusing equation becomes more complex.

9.1.3 Acoustic Holography

Detection of the reflection of the formed beam impinging on the target area is carried out by means of a *holographic reconstruction technique*. This relies on the detection of both amplitude and phase information contained in the

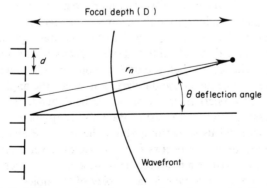

Fig. 9.5. Effects of simultaneous focusing and deflection.

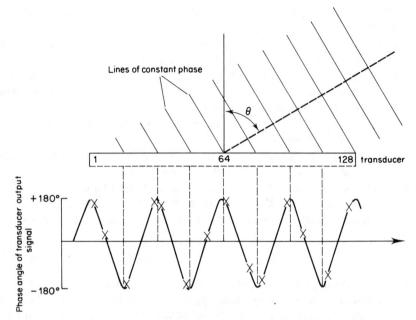

Fig. 9.6. Instantaneous waveform along receiving transducer.

reflected signals available at the receiver array. A brief description is given below; more detailed information on the theory of acoustic holography can be found in [3].

Consider Fig. 9.6. Sound incident upon the array from a distant point reflector will be in the form of plane waves. In Fig. 9.6 these waves are depicted by straight parallel lines at an angle θ to the normal array. Between the lines of constant phase shown on the diagram the phase of the signals will change linearly with distance. If we consider a particular instant of time the phase of the signals incident upon the array will change sinusoidally along the length of the array. The number of cycles of sinusoidal variation will depend upon the angle of incidence, θ. If $\theta = 0$ (i.e. arrival of the sound waves along a line normal to the array) then the lines of constant phase will lie along the array and there will be no sinusoidal variation. Since the angle of incidence θ for a point source reflection can be shown to be dependent upon the spatial frequency measured across the array then the frequency spectrum of this measurement will tell us something about the multiple angles of incidence present whenever reflection occurs from an irregular solid object (Fig. 9.7). What we actually derive is a holographic representation which can be processed further to provide a conventional display of the object.

Once again, when objects are close to the array, the situation becomes

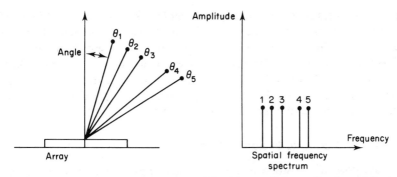

Fig. 9.7. Showing multiple angles of incidence from reflective object and the resulting frequency spectrum.

complex. Figure 9.8 shows that in the near field of an array the sinusoidal wave becomes distorted and has the appearance of a swept frequency. Such a signal, when transformed into a frequency plane would give rise to many components, even for a single point source. This results in a "blurring" of the image and, as with the phased array used for transmission, it is desirable to focus the effective object source of the received signals by slight adjustments to the phase shifts of the signals present at the holographic receiving array elements. This is necessary particularly to correct the distortion which would be apparent in near objects due to the curvature of the incoming wavefronts [3].

Instead of carrying this out directly it is advantageous to modify the digital form of the signals. If the signal from each transducer is converted to two

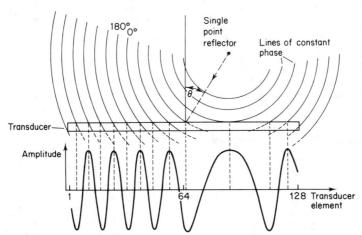

Fig. 9.8. Waveform for a point in the near field.

digital numbers representing amplitude and phase, then it is possible to add a phase correction to the phase term associated with each element. This correction acts mathematically to "straighten" the wavefronts coming from an object in the near field and thus removes the associated distortion. Again a repetitive phase correction is required for each element and needs to be carried out for any depth of field. This forms a task for the microprocessor associated with the holographic receiver unit.

9.1.4 An Acoustic Imaging System

We are now in a position to describe the complete system, is based on the work carried out at Newcastle University [4]. A block schematic diagram is given in Fig. 9.9. Three microprocessors are employed. Of these one is a

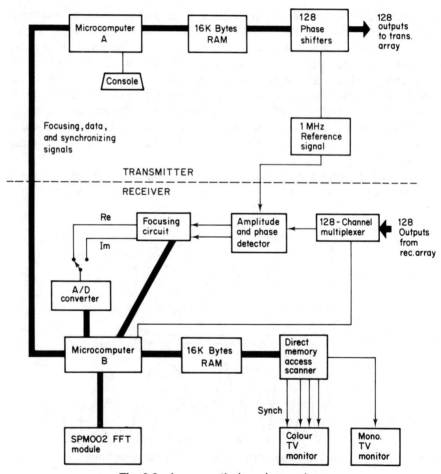

Fig. 9.9. An acoustic imaging system.

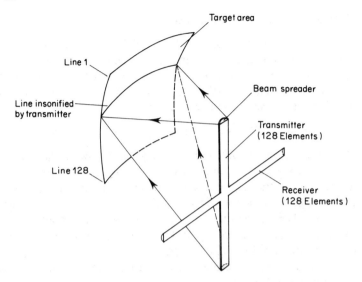

Fig. 9.10. Transducer arrays and scan format.

dedicated fast Fourier transform processor based on the Plessay "Miproc" high-speed microprocessor. The other microprocessors are Motorola MC6800 devices contained in microcomputers A and B of Fig. 9.9. The two microcomputers are identical and comprise CPU, 32K RAM, disc drive, high-speed A/D and D/A converters, parallel and RS232 interfaces.

Microcomputer A controls the bank of 128 phase-shifting units via an additional 16K bytes of RAM. Microcomputer B controls the acquisition, transformation and display of the final image obtained from the array of 128 receiver-transducers. Figure 9.10 illustrates the scanning procedure adopted. The 128-element phased array transmitter is arranged vertically to produce a wedge-shaped beam of sound focused at the desired depth of observation. This has the effect of a thin line on the surface of an imaginary hemisphere whose centre lies at the intersection of the vertical transmitter and horizontal receiver arrays. Thus all points along this line will receive acoustic energy. Any reflective points will redirect a portion of the incident energy back to the receiver array. Such points will then give rise to a spatial frequency spectrum of shape determined by the angular positions of the reflecting points from the array (Fig. 9.7). It is the task of microcomputer B to compute the angular positions of these reflective points and to display them on the CRO. The scanning arrangement for the display CRO is identical to that found in the television raster but with the number of lines limited to 128. As the transmitter beam pauses at the end of each scanning line, sufficient time is allowed for the receiver to sample the amplitude and phase of the signal present at each array element and to compute the resulting image.

9.1.5 The Phased Array Transmitter

Conversion of the calculated value of β (Eq. 9.1) into an actual shift in the phase of the signal applied to each transmitter element is achieved by a *phase shifter* circuit, one for each element. The phase shifters are driven from a common 1 MHz reference signal input which is modified for each separate element in accordance with the relationship:

$$\sin(\omega t + \beta) = \sin(\beta)\cos(\omega t) + \cos(\beta)\sin(\omega t) \qquad (9.2)$$

where β = desired phase shift and ω = angular frequency of the acoustic signal. The electronic realization of this equation is shown in Fig. 9.11. This circuit is driven from the reference sine wave and multiplies this by $\cos\beta$ and $\sin\beta$. After phase shifting one of the multiplier outputs by 90° (to produce $\cos\omega t$), the signals are added to obtain the phase shifted output. The multiplication is carried out by *multiplying D/A converters* (MDAC), which allows an analog output to be produced from separate analog and digital inputs. The analog input is the reference input; the digital input represents the cosine or sine of the desired phase angle β.

The circuit diagram of a single phase shifter is shown in Fig. 9.12. The interface is via a pair of eight-bit latches into which are fed the cosine or sine of the desired phase angle, previously stored in ROM. Since the scan circuits require an appreciable time to initiate all 128 pairs of values (these must be set up sequentially), a further set of latches (the transfer latches) are used to ensure that all the phase shifters move to their new values simultaneously.

The overall system diagram for the transmitter is shown in Fig. 9.13. Control of the phase shifters is shared between the DMA scanner and the microprocessor. It is a task of the microprocessor to calculate and store (in the four banks of 4K byte RAM) all the phase values needed for a complete scan. This implies a total of 128×128 (16K) values to set 128 phase shifts for each of the 128 scan positions. These calculations are performed only when it is required to re-focus the system at a new depth. It can take several seconds to evaluate the value of β in Eq. (9.1) for each of the 16K values needed.

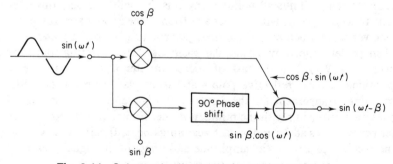

Fig. 9.11. Schematic diagram of a phase shifter.

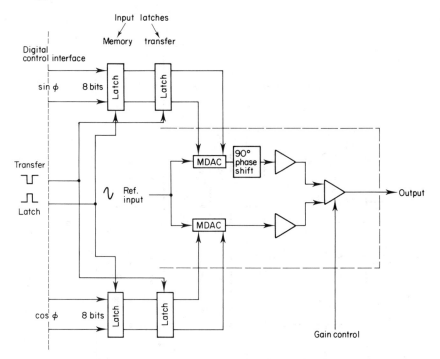

Fig. 9.12. System diagram of a phase shifter.

Once a scan operation is commenced, however, the microprocessor would not be able to respond sufficiently rapidly to process β values at the rate of one every 0·5 μs (i.e. 128 values in one scan period of 64 μs). Instead, the data stored in RAM is retrieved and supplied to the phase shifters via the fast TTL logic of the DMA interface (see Section 5.3).

The phase shifters are arranged in groups of 32 and each of these groups is serviced by 64K bytes of RAM. In order to avoid unnecessary use of RAM, only the desired phase angle is computed and stored. To produce cosine and sine values from the stored data, two look-up tables (ROM store) are used per bank of 32 elements.

Some degree of amplitude control was found necessary to allow the use of focusing in the near field of the array [3]. Amplitude information is stored in an additional area of RAM.

9.1.6 The Holographic Receiver

Referring to Fig. 9.6, the signals from the receiver array are routed via a 128-way multiplexer, controlled from microprocessor B, to an amplitude and phase detector. The microprocessor selects each transducer in turn and

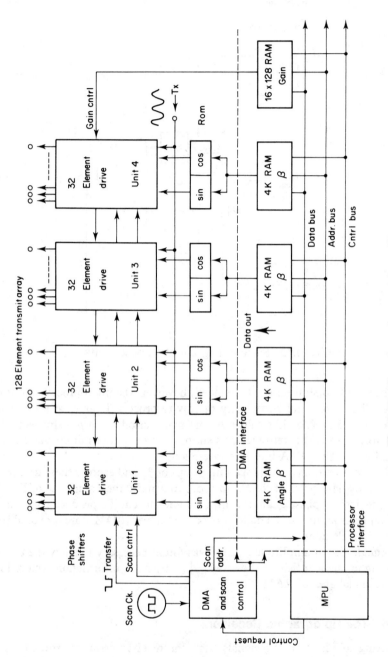

Fig. 9.13. The phased array transmitter.

measures its output signal in terms of amplitude and phase. As discussed earlier, this permits reconstruction of an image by means of Fourier transformation.

The amplitude and phase detector produces two voltages at its outputs representing the real and imaginary parts of the complex signal. If we let

$$V = A \exp [j(\omega t + \phi)] \qquad (9.3)$$

be the signal, then this will produce two outputs:

$$\text{Re} = A \cos \phi \qquad (9.4)$$

and

$$\text{Im} = A \sin \phi \qquad (9.5)$$

Focusing of the holographic receiver, as discussed in Section 9.1.3, is obtained by multiplication. The operation to be carried out is to correct the phase value inherent in the value of Eq. (9.3) by subtracting an amount $\Delta\phi$. This can be expressed as the multiplication of V in Eq. (9.3) by another complex signal, $\exp(-j\omega\Delta\phi)$, i.e.

$$V' = A \exp [j(\omega t + \phi)] \exp (-j\omega \, \Delta\phi) \qquad (9.6)$$

Expressing Eq. (9.6) in terms of real and imaginary components this yields:

$$\text{Re}' = A(\cos \phi \cos \Delta\phi - \sin \phi \sin \Delta\phi) \qquad (9.7)$$

$$\text{Im}' = A(\sin \phi \cos \Delta\phi + \cos \phi \sin \Delta\phi) \qquad (9.8)$$

where Re$'$ and Im$'$ are the real and imaginary focused outputs.

The arrangement used to achieve focusing is shown in Fig. 9.14. This is carried out entirely by analog circuits in which the unfocused outputs of the detector elements are fed into four analog multipliers which also receive the focusing terms ($\sin \Delta\phi$ and $\cos \Delta\phi$) supplied from microcomputer B via D/A converters. As with microcomputer A, the receiver focusing data is computed once for a given depth of examination. As the scanning plane is located at a constant depth from the receiver, however, only 128 values need to be calculated by microcomputer B. Thus the steps taken by microcomputer B to form one sample from one element are

(1) Select the correct element by supplying its address to the multiplexer.
(2) Feed the cosine and sine of the focusing correction factor to the focusing circuit.
(3) By means of a two-channel multiplexer obtain the focused real and imaginary terms and store them.

These steps must be repeated 128 times in order to complete one scan line.

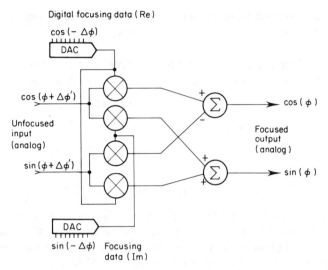

Fig. 9.14. The focusing circuit.

The input data acquired in this way (128 complex numbers) form the input data to the fast Fourier transform unit.

This is a Plessey SPM002 unit within which the FFT transformation of 128 complex input values is carried out by software implementation of the Cooley–Tukey algorithm to achieve a complete transformation and ordering of the output terms in 22 ms. The output is presented as a spectral density (i.e. squared FFT terms (see Section 7.5.1)) in order to indicate the total energy arriving from a particular angular direction.

Having submitted the data to the SPM002 and obtained the spectral density values at its output, microcomputer B stores the resulting image line in an area of RAM. The RAM is accessed by a purpose-built display unit which displays the data stored in 16K bytes of RAM as a 128×128 point image on the television screen. The image is available in grey scale or pseudo-colour form (Fig. 9.9).

It will be appreciated that the output of the FFT unit actually gives angular information about the object under examination (Fig. 9.7). A further task of microcomputer B is to convert the polar image coordinates into Cartesian coordinates so that the actual shape of the object can be realized.

Communication between microcomputers A and B is provided through a parallel interface. This allows synchronization to be achieved between the receiver and transmitter scans.

In this particular example, the limitation in speed of software FFT conversion precludes operation in real time, which is desirable for this type of application. Later developments of this body scanning technique use faster methods of frequency conversion (discussed in Chapter 7) such as hardware conversion using fast charge-coupled devices (CCD) [5].

9.1.7 System Control

Control of the complete imaging system is illustrated in Fig. 9.15. This commences with overall system reset at which time microcomputer A takes control of the system. Initialization begins with reading-in scan parameters from the operators console. The data requested by the system are depth of focus and scan angle.

The parameters are entered into the memory of microcomputer A which relays the parameters also to microcomputer B. Both computers commence calculating their respective focusing terms, described earlier, and microcomputer A also calculates the adjusted phase shifts required for each transmitting array element. At the same time microcomputer B initiates set-up values required for the FFT module and transfers these to the third microcomputer contained in the FFT module so that it is in a positon to receive input data.

Pulse sound transmission commences at array position 1 under the control of microcomputer A. This continues sequentially for each array position with a check for operator interrupt between each pulse transmission. A short delay is inserted in the controlling program for microcomputer B before this accepts inputs from the receiver array and processes these values for transmission to the FFT processor.

During the intervals between transmission of sound pulses from each array position the FFT is computed and values are sent to microcomputer B for further processing and coordinate conversion. Display of the first line of the reflected image occurs before the second position is reached in the control of transmitted sound pulses by microcomputer A.

It will be seen that many concurrent tasks are being performed by the two microcomputers A and B and this is therefore a good example of the kind of time-sharing operation described in earlier chapters when we dealt with system control.

9.2 DATA LOGGING OF SOLAR ACTIVITY THROUGH RADIO WAVE ABSORPTION

Our second example is concerned with recording data in an on-line environment. Although only a single channel of data is involved, the stringent requirements for data collection and the range of facilities needed in field data logging equipment makes this a good example of the economy which can be achieved in the use of a microprocessor to control the various processes involved.

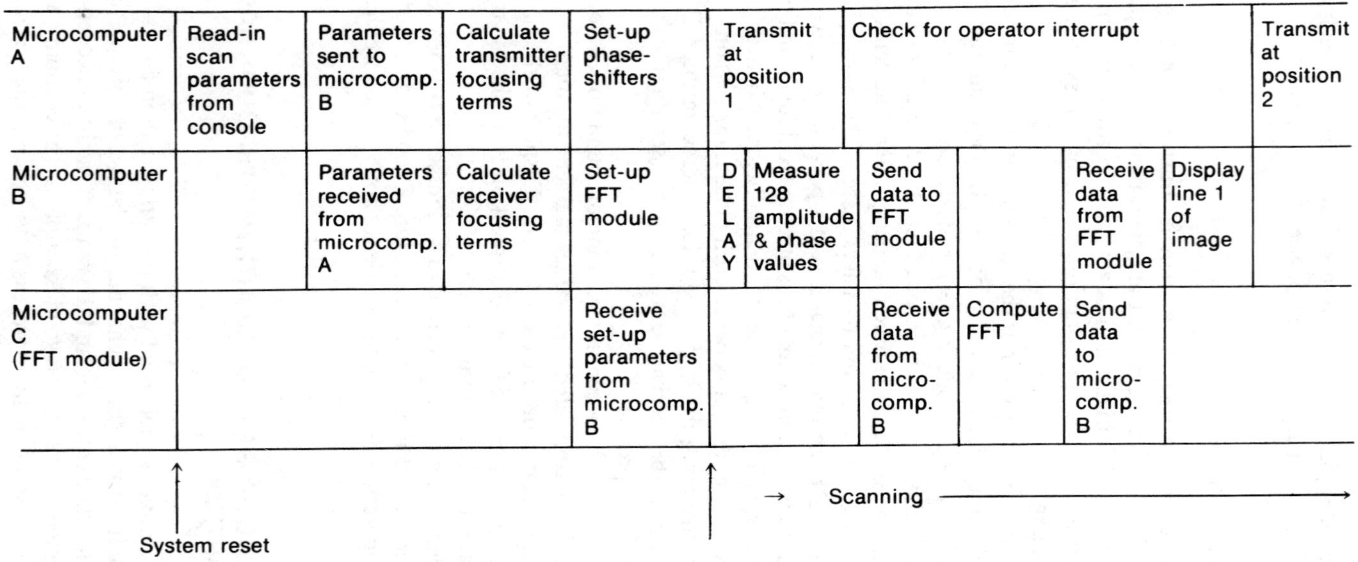

Microcomputer A	Read-in scan parameters from console	Parameters sent to microcomp. B	Calculate transmitter focusing terms	Set-up phase-shifters	Transmit at position 1	Check for operator interrupt			Transmit at position 2
Microcomputer B		Parameters received from microcomp. A	Calculate receiver focusing terms	Set-up FFT module	D E L A Y — Measure 128 amplitude & phase values	Send data to FFT module		Receive data from FFT module	Display line 1 of image
Microcomputer C (FFT module)				Receive set-up parameters from microcomp. B		Receive data from micro-comp. B	Compute FFT	Send data to micro-comp. B	

System reset ← → Scanning

Fig. 9.15. Timing control of the complete system.

9.2.1 Measurement of Radio Wave Absorption in the Ionosphere

Measurement here concerns the absorption of extra-terrestrial radio waves in the upper atmosphere which at high latitudes is related to the incidence of energetic protons or auroral electrons and in its turn is a consequence of enhanced solar activity. The record obtained is in the form of a varying voltage over a 24-hour period which may be compared with other records obtained at other times to ascertain the level of absorption. One transducer employed by radio scientists to measure ionospheric absorption is known as a riometer [7]. Such a device produces an output voltage proportional to the intensity of cosmic radio noise received at any instant. A limited amount of information can be derived from visual examination of the time history of a pen recorder trace, but, for reasons discussed in Chapter 7, it is advantageous to undertake some microprocessor processing and control of the data logging operation.

The signal from the riometer contains not only useful information but also background noise. Normally this noise is rejected during interpretation by averaging readings taken from the record. A typical record of 24 hours activity is shown in Fig. 9.16. To obtain the actual level of activity at any particular time, use is made of a *quiet day curve* (Fig. 9.17). This is a record of the reference level or background noise present on a day of minimum solar activity. The true level is obtained from the ratio of these two records.

Derivation of this ratio is normally carried out some time after the record is

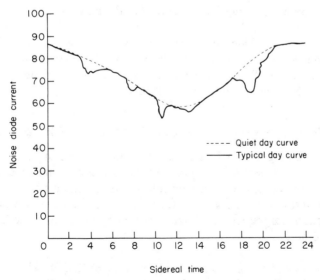

Fig. 9.16. A 24-hour record of sunspot activity.

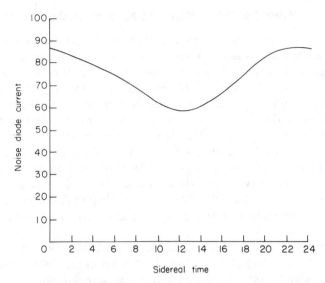

Fig. 9.17. A quiet day record for sunspot activity.

obtained, since recording installations are often located in isolated regions and the record sent to a base station for interpretation.

It is, however, desirable to derive a real-time indication of the ratio absorption and to arrange for the recording riometer to read quiet day information from interval storage and to carry out the necessary signal averaging and noise rejection. At the suggestion of Dr J. K. Hargreaves of the University of Lancaster, consideration has been given to how the operation of a riometer and the subsequent data processing may be aided through microprocessor techniques. What follows is a discussion of the design principles underlying the development of such a logger.

9.2.2 A Microprocessor-based Data Logger for a Riometer Unit

A schematic diagram of the data logging unit is given in Fig. 9.18. The unit consists of five sections carrying out the functions of (a) Processor and Storage, (b) Timing and Failure Control, (c) Digital/Analog conversion, (d) Function Control and Display, and (e) Data storage and Disc Control.

These individual sections are interconnected by means of an address data and control bus.

(a) *Processor and storage* This card provides a 6502 CPU with 12K bytes of ROM and buffer registers on both address and data lines. Buffers are

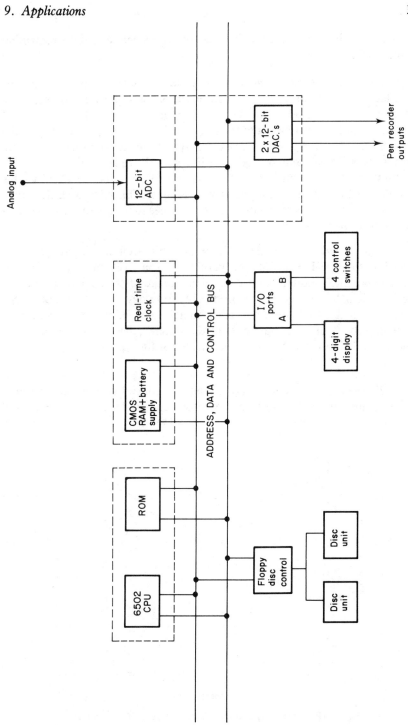

Fig. 9.18. The data logging unit.

required in order to provide the drive capability needed to feed the remaining modules of the system.

(b) *Timing and failure control*   The system is mains-powered and un-attended. To avoid loss of data and timing signals, a standby battery supply is automatically connected in the event of mains failure. In the RAM are stored various control words relating to the activity which was being recorded at the time of power failure. This allows the system to avoid overwriting data already recorded and to mark the recording with the time of interruption and its duration. The system is provided with a real-time clock, located in this module, giving day, hours, minutes and seconds. This CMOS clock chip is also connected to the battery circuit and continues to provide accurate time in the event of mains failure.

(c) *Digital/analog conversion*   Although 8-bit accuracy is adequate for recording the signal over a 24 hour period, a 12-bit A/D conversion is neces-sary to take into account the wide dynamic range of the signal plus a mean d.c. level which can vary quite considerably from day to day (see Fig. 9.19).

The two 12-bit D/A converters allow signals to be fed to a pen recorder for compatibility with other non-automated systems. Two channels are pro-vided, one being the raw data, the other recording the accumulating quiet day curve.

(d) *Function control and display*   Function control of the data logger is obtained through the setting of four panel switches and indicated by means of

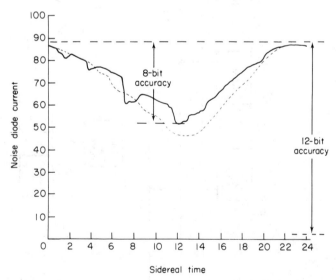

Fig. 9.19. Showing the dynamic range of sunspot activity.

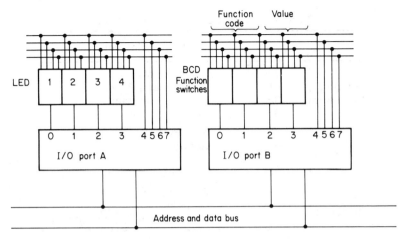

Fig. 9.20. Function control and display.

a four-digit LED display. Two of the switches set a function code which determines the mode of operation of the data logger. The other two switches are for the insertion of numerical parameters (if required).

A single 6522 Versatile Interface Adapter (VIA) is used to connect the four-digit display and four BCD control switches to the I/O ports. The switches produce the binary equivalent of the decimal number selected. The binary outputs are connected to a common line via diode OR gates and connected to bits 4–7 of port B on the 6522. The other four bits of port B are used to place a logic 1 on one of the BCD switch "common" inputs and logic 0 on all the others. This allows each switch to be addressed in turn and the binary value corresponding to the switch setting to be transferred to the BCD lines.

The interrogation of the switches is performed under software control. The 6502 CPU sets up port B such that bits 0 to 3 are selected as outputs and bits 4 to 7 as inputs. This is achieved by the CPU writing hexadecimal value 0F into the data direction register which configures port B on the 6522 VIA chip. To read the setting on, say, the extreme left-hand switch in Fig. 9.20, the CPU first writes the value 01 hex to port B which places a logic 1 on bit 0 and logic 0s on bits 1, 2 and 3.

The CPU can then read the switch value from port B. Note that this value will appear in the upper nibble of the accumulator if the latter is used to read from port B. The operator indicates that a function has been requested by depressing an "enter" button thus initiating an interrupt to the CPU when a logic 1 is present.

The following function codes may be set by the switches together with a two-digit number if required by the function.

Table 9.1. Data logging function codes.

	Code	Meaning
Logging functions	00	Select display†
	01	Commence logging
	02	Terminate logging
	03	Set sample rate†
Clock set	04	Set day†
	05	Set hours†
	06	Set minutes†
	07	Set seconds†
Disc functions	08	Format disc†
	09	Verify disc
	10	Report free disc space (Kbytes)

†Requires value to be supplied.

Most of the above functions are self-explanatory. the "select display" function requires a value to be supplied to select one of the following:

Table 9.2. Select display function code.

Value	Display
00	Raw
01	Averaged data
02	Absorption in dB.
03	Time (hours, minutes)

The "format disc" function requires a value to be entered as a precaution to reduce the risk of accidental erasure. The operator maintains a log of such values, for disc identification. "Verify disc" allows the operator to check the disc for correct formatting and act to discard the disc if a flashing 0000 indication shows that the disc is faulty.

The display in this module is a seven-segment liquid crystal described in Chapter 3. This includes BCD-to-seven-segment decoders arranged to latch information fed to them on receipt of a display flag—a technique which we noted earlier allows commoning of several BCD inputs.

The display BCD inputs derive from port A of the VIA. Bits 4 to 7 are used to carry the BCD information and bits 0 to 3 the flags. The CPU writes FF hex to the data direction register for port A to select all bits as outputs. To update a given display the CPU first places the received BCD values on bits 4 to 7 with bits 0 to 3 all set to logic 0. This is repeated but with the required flag line (say, bit 0) set to a logic 1. Finally the first step is repeated and all 0s are written to the lower four bits (keeping the data on bits 4 to 7). This procedure is necessary in order to flag the BCD information correctly into the BCD-to-seven-segment decoder. The sequence is repeated four times with appropriate BCD data being entered into each display. The use of software control of the display allows it to be used for the variety of different functions given by Table 9.1.

(e) *Data storage and disc control*   This card contains a Western Digital FD 1791 floppy disc controller and its support circuits. Two $5\frac{1}{4}$-in. floppy disc drives are used in double-density recording format. This permits over 200 Kbytes of information to be stored per disc. One disc drive is dedicated to logging new data and the other used to retain the quiet day record.

Software routines control reading and writing of data from and to the discs. Other routines provide formatting and verification control.

We mentioned earlier that a quiet day record may be obtained of readings at given sidereal time taken over several days. Such a quiet day record is built up on one disc by adding-in each day's reading at the end of the day and normalizing the sum. The new quite day record is then used to obtain the actual absorption value on the following day. This value is the ratio between a given reading and a corresponding reading on the appropriate section of the quiet day record.

The software rejects large noise transients present in the raw data by continually looking at the difference in level of successive samples. If this difference exceeds a given threshold value then the sample is not displayed, although *all* data is logged so that subsequent analysis can take into account all information obtained.

As each sample is taken it is recorded in *packed* form together with a 12-bit quantity signifying the time (in units of $\frac{1}{2}$-second) since the last sample was taken. Thus each sample occupies three 8-bit bytes on the disc (Fig. 9.21). The fastest sampling rate is two samples per second and the slowest six samples per hour. At the start of each run a header is recorded which specifies the sampling rate chosen for the day together with the time at which recording commenced.

It may be considered unnecessary to store the time at which each sample is chosen if the sampling rate is known. However, one of the sampling rates allows for sampling to occur only when the signal has changed by a given amount. This sampling method can reduce the amount of data stored and thus increase the available logging time per disc. To sample only when necessary leads to more efficient disc utilization. However, it does imply a variable sampling rate and this needs to be recorded.

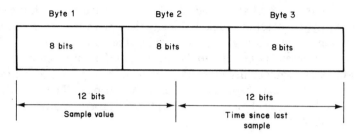

Fig. 9.21. Data packing in a disc record.

Integrity of data is important and is ensured by reading back each block of data as it is stored to check that it has been written correctly. If incorrect data is found, then up to ten attempts are made automatically by the system to rewrite the data, reading back each time. Should ten unsuccessful attempts be made then the CPU is arranged to mark that area of the disc as unusable and moves to another free area. A record of faulty sectors is retained on the first track of the disc.

9.2.3 The Replay Unit

Two possibilities exist for replaying recorded information. One is via a conventional microcomputer which shares the same floppy disc recording format and is linked to some display or hard-copy read-out device. This approach is inflexible since it is dependent on the availability of the correct disc recording format on the disc drive. An alternative method implemented with this system is to construct a special replay unit to the appropriate standard. This is similar to the data logger but contains only the CPU, ROM, RAM and floppy disc controller.

The software of the replay unit is written to transfer data from disc to either a printer or a linked computer via the parallel interface available at the Function Control card which acts as an I/O route. This allows either visual inspection of the data or signal processing using the linked computer. The analysis results may then be plotted by a graphics device attached to the computer and advantage can be taken of the statistical and other program packages usually available on the larger machine.

9.2.4 Summary

We can see from this description that microprocessor control of even a single data channel can confer a number of operational advantages. Real-time signal processing and analysis may be carried out to permit display of the actual level of absorption. Recording of this data together with accurate time information may be arranged to occur at varying sampling rates and, for example, only when a given threshold of activity is exceeded. The recording medium employed (floppy discs in our example) permits rapid acquisition of data, a high packing density and ability to transfer the recorded data to other replay equipment associated with perhaps more complex processing computers. Data transmission control in a computer network environment is a further advantage of microprocessor control of such acquisition methods and this applies to many of the process control methods described in this book. Finally, two operational advantages of value in field investigations must be mentioned. These are the flexibility conferred in system design which enables even complex recording, processing and calibration procedures to be carried out by relatively unskilled personnel and the ease with which "house-keeping" information related to the signal (record number, time, sampling rate, calibration constants, etc.) may be added to the recorded data.

References

1. Robinson, G. P. S. (1978). Acoustic Imaging. *Ph. D. Thesis*, University of Newcastle upon Tyne, U.K., 39–61.
2. Hildebrand, B. P. and Boenden, B. B. (1977). "An Introduction to Acoustical Holography". Plenum Press, New York.
3. Keating, P. N., Koppelman, R. F. and Mueller, R. K. (1974). Complex on-axis holograms and reconstruction without conjugate images, *in* "Acoustic Holography", vol. 5, pp. 515–526. Plenum Press, New York.
4. Posso, S. M., Kennair, J. T. Robinson, G. P. S. and Taylor, B. A Microprocessor-based Ultrasonic Phased Array Scanner (to be published in *Signal Processing*, North Holland).
5. Buss, D. D., Veenvant, R. L. and Broders, P. L. (1975). Comparison between the CCD chirp Z Transform and the digital Fast Fourier Transform, *Proc. Int. Conf. on Applications of Charge Coupled Devices*. San Diego, California, Oct.
6. Little, C. G. and Leinback, H. (1958). Some measurements of high altitude inonospheric absorption using extraterrestrial radio waves, *Proc. IRE* **46**, 336–48.
7. Hargreaves, J. K. (1969). Auroral absorption of H. F. radio waves in the ionosphere, *Proc. IEEE* 57(8) 1348–1373.

Index

Absolute addressing, 114
Accumulator, 109
  manipulation instruction, 130
Acoustic holography, 287
  imaging, 282
  imaging system, 290
Ada compiler, 197
Adder design, 80
  full 52
  floating point 91
  half 46
Address, 9
  base, 28
  deferred, 116
  direct, 114
  field, 126
  immediate, 114
  information, 9
  indexed, 118
  indirect, 116, 127
  indirect machine, 28
  mode, 114, 112
  pins, 12
  pointer, 145
  relative, 119
  return, 124
Advanced Micro Devices Ltd., 245, 274
Aliassing filter, 250
ALGOL, 199
Analog-digital conversion, 21, 249, 255, 258
Analog I/O connection, 262
  I/O board unit, 263
  interfaces, 21
AND gate, 43
ANSI Y32.14(1973), 45
Applications, 282
  program, 32

Arithmetic and logical instruction, 126
A-scan system, 283
ASCII Code, 17
Assembler, 136, 196
  Intel 8080, 137, 162
  language, 136
  programing, 136
  two-pass, 137
Autocorrelation, 234
Autocorrelogram, 234
Axioms–Boolean, 48

Base address, 28
BASIC interpreter, 198
Bell Laboratory DSP, 244
Bi-directional data bus, 157
Binary-weighter ladder, 254
Bistable (see Flip-flop)
Bit-by-bit correlation, 237
Bi-polar transistor, 69
Bit-reversed order, 228
Bit slicing, 103
  architecture, 215, 231
  arithmetic, 232
  module, 71,
Boolean algebra, 47
  arithmetic module, 76
  axioms, 48
  expression, 48
  logic design, 36
  sequential module, 84
  synthesis process, 72
Booth's algorithm, 279
Branching, 118
  conditional, 118
  unconditional, 118
Breakpoint, 207

B scan method, 284
Buffering register, 16
Bus, 57
  bi-directional, 157
  IEEE 488 standard, 187
  Intel multibus, 190
  width, 9
Butterfly module, 233
  operation, 227
Byte, 9

Capacitor charging method, 256
Capacitor ROM, 96
Carry bit, 139
  look ahead adder, 78
Charge coupled device, 297
Chip connection, 9
Clock-controlled flip-flop, 65
Clock rate, 155
CMOS (Complimentary MOS), 69
  construction, 250
COBOL, 199
Code, 17
  condition, 118
  function, 304
  ASCII, 17
  pseudo, 137
Command language, 223
Communication interfaces, 21
Comparator, 81
  5-bit, 81
Complement, 41
Compiler, 148, 199
  ada, 198
  algol, 199
  cobol, 199
  forth, 199
  fortran, 199
  pascal, 149, 199
Condition branch, 118
Condition codes, 118
  register, 111
Context-dependent machine instruction,
  151
Context switching, 121
Control and status register (CSR), 263
Control information, 9
Convolution, 97
Cooley and Tukey algorithm, 225
Correlation (see also Autocorrelation), 234

bit-by-bit, 237
  polarity coincidence, 239
  using the FFT, 235
Correlator chip, 238
Cosine Fourier transform, 236
Counter, 87
  up, 87
  up down, 88
CP/M operating system, 149, 197, 202
Cross assembler, 197

Daisywheel devices, 16
Data burst, 222
  communication equipment, 187
  logging, 218, 271, 297
  management, 30
  movement instruction, 127
  sampling, 222
  terminal equipment, 187
Deadlock, 31
Debugging, 183, 195, 207
Decimal adder, 76
  arithmetic correction, 77
Decoder, 56
Deferred addressing, 116
De Morgan's theorem, 50
Demultiplexing, 56
Development systems, 194
Device handlers, 25
  interfaces, 185
  management routine, 31
  priorities, 168
Diagnostic medicine, 283
Digital-analog converter, 21, 253, 302
  correlation process, 237
  filters, 98, 266, 268
  signal processor, 244
Diode-transistor logic (DTL), 71
Direct addressing, 114
Direct memory access (DMA), 183
  address, 184
  Intel, 185
  transfers, 222
Disable interrupt instruction, 170
Disc block directory, 20
  drive unit, 17
  floppy, 17
Discrete Fourier transform, 224
Documentation, 39, 196
Domain transformation, 217

Dot matrix, 16
Double-byte numbers, 146
Double-precision working, 267
Dynamic memory cells, 95
   memory management, 27
   vector, 123

Edge-triggering, 67
Editor program, 202
Electronic switch, 41
Emittor-coupled logic (ECL), 71
Emulation – in circuit, 207
Enable interrupt, 170, 172
Erasable PROM, 11
Error handling routine, 32
Evaluation kit, 197
Exclusive-OR gate, 46
External reference, 137

Falling edge, 67
Fan-in, 51
Fast Fourier transform (FFT), 99, 201
FFT algorithm, 224, 296
   processor, 230
Field effect transistor, 69
Files, 202
Filter, 266
   aliassing, 250
   digital, 266
   FIR, 267, 278
   IIR, 269
   non-recursive, 267, 278
   performance, 267
   recursive, 268
Finite impulse response filter (see Filter
   FIR)
Flip-flop (see also Status latches), 63
   clock controlled, 65
   J.K., 65
   set–reset, 63
Floating point adder, 91
Floppy disc, 17
Flow chart, 196
FORTH, 199
FORTRAN, 199
   77, 200
   80, 200
Fourier transform (see FFT)
Full adder, 52

Full duplex, 187
Function code, 304

General purpose register, 104

Haar orthogonal transform, 230
Half-adder, 46
Half-duplex, 187
Hardware, 8
   design aids, 205
   development, 35
   development aids, 204
   testing, 207
HELP facility, 221
Hewlett Packard HP49000
   Microprocessor, 208
High level language, 38, 147, 198
High speed memory cells, 13
Holographic receiver, 293
   reconstruction, 285
House-keeping information, 307

IEEE-488 bus standard, 187
Image processing systems, 217
Immediate addressing, 114
Immediate mode move instruction, 127
In-circuit emulation, 207
Indexed addressing, 118
Indirect addressing, 116, 127
Infinite impulse response filter (see Filter
   IIR)
Information transfer, 16
In-place computation, 228
Input/Output (I/O) devices, 15
   device driver, 167
   device handler, 153, 167
   hardware, 155
   ports, 164
   programming, 161
   request, 154
   system, 153
   subsystem, 3
Instruction, 128, 170
   arithmetic and logical, 126
   decode, 6
   disable interrupt, 170
   execution process, 13
   fetch, 6

fetch cycle, 14
jump, 139
load immediate, 128
long jump, 119
register, 109
set, 4
Integrated circuit, 8
large scale, 68, 76, 93
medium scale, 215
very large scale, 215
Integrated injection logic, 71
Intel 4004 Microprocessor, 103
8008 Microprocessor, 103
8080 Microprocessor, 109, 198
8080 assembly language, 137, 162
8080 machine instructions, 125
8085 Microprocessor, 100, 104, 164
8086 Microprocessor, 30, 108, 150
8357 programmable DMA controller, 185
multibus, 190
SDK 85 kit, 205
Interconnected microprocessors, 243
International standard, 45
Interface adaptor, 161
Interfacing, 3, 165
standards, 186
Interpreter, 198
Interrupt, 26
acknowledge signal, 169
control unit, 174
enable, 173
function, 25
handling, 169
high priority, 168
operation, 263
programming, 178
requests, 168
servicing routine, 179
source recognition, 167
structure of Motorola 68 000, 170
systems, 166
vectoring, 168
Interval timer, 26
Inverse discrete Fourier transform (see also FFT), 224
Inverter, 42

J-K flip-flop, 65
Jump instruction, 118, 126, 139

long-instruction, 120
Junction transistors, 69

Karnaugh map, 55, 75
Keyboard, 163
selection, 221
Keywords, 198

Lancaster University, 300
Large memory module, 93
Large scale integration, 68, 76, 93
Last-in-first-out structure, 123
Latency time, 20
LED display, 38, 205
Line sharing, 190
Load immediate instruction, 128
Logic circuits from NAND/NOR gate, 58
devices, 52
diagram, 44
families, 71
gates, 41
state analysers, 209
Logical design, 156
of Boolean devices, 72
Long jump instruction, 120
Look-ahead fetching, 244
Loop, 118

Machine address, 28
Macros, 140
Maintenance, 39
Matrix multiplication, 98, 224
MDAS data logger, 219
Medium-scale integration, 215
Memory, 9
cell, 94
cell – high speed, 13
control lines, 12
erasable programmable (EPROM), 11
management, 26
mapped I/O, 23
overlays, 27
programmable (PROM), 11
random access (RAM), 95, 164
read only (ROM), 11
segmentation, 30, 107
subsystem, 3
word, 9
Microelectronics, 215

Microprocessors, 100
  applications, 3
  architecture development, 149
  programming, 103
  system, 12
  systems development, 34
Microsoft BASIC, 198
MOS (Metal Oxide Silicon), 69
  bi-polar transistors, 250
  technology, 267
  technology 6502 processor, 108
MOSTEK, 258
Motorola 6800 microprocessor, 104, 198,
  219, 291
  68 000 microprocessor, 150
M-tap filter, 280
Multi-device handling routines, 183
Multiplexing, 21, 56, 259
Multiple processing units, 6
Multiply–add function, 243
Multi-programing, 121
Multi-tasking, 121
Multi-user disc file, 202

NAND gate configuration, 51, 59
Newcastle University, 290
  MOS transistor, 69
Non-recursive filtering, 267, 278
NOR gate, 51
N-p-n transistors, 69

Opcode, 126
Operational amplifier, 250
Operating systems, 202
Order of a filter, 269
OR gate, 42

Packing density, 307
Packed form, 305
Paging system, 30
Parallel arithmetic operation, 279
  methods, 273
  multiple bus architecture, 245
  processing architecture, 276
  sorting, 99
  vector processing, 97
PASCAL, 199
  UCSD, 149, 199

P code, 199
Peripheral interface adapter (PIA), 161,
  271
Personality modules, 209
Phased array, 285
  transmitter, 292
Phase shifter circuits, 292
Pipe-line process, 244, 267
Plessey Miproc high speed
  microprocessor, 291
PMOS family, 71
P-n-p transistor, 69
Pointer manipulation instruction, 126
Polarity coincidence correlation, 239
Polling, 168
Popping, 125
Power spectral density, 236
Priority encoder unit, 173
Problem oriented language, 5
Processor, 6
  management function, 25
Programmable interrupt controller, 178
  Signal Processor, 218, 241
Program counter, 111
PROM (Programmable read-only
  memory), 11
Prototyping kit, 205
Pseudo code, 137
Pushing, 125

Quantizing, 250

Ramp and hold converter, 255
Random-access memory (RAM), 95, 164
Read-only memory (ROM), 11
Real address, 28
Recursive filtering, 246, 268
Register, 13, 84
  addressing, 127
  buffering, 16
  instruction, 109
  manipulation, 135
  shift, 85, 96
Relative addressing, 119
Resistor-transistor logic (RTL), 71
Return address, 124
Riomer, 299
Ripple-adder, 52
ROM modules, 11, 95

Run-time package, 199

Sample and hold amplifier, 251
Sampling, 250
Scanning system, 284
Screen editor, 203
Semiconductor storage cells, 203
Sequential circuits, 63
  filter, 276
  multiplier, 89
Set–reset flip-flop, 63
Shift-register, 85, 96
Shuffling, 228
Signal flow graph, 227
  processing, 215
  processing language, 246
Simulators, 201
Sine Fourier transform, 236
Single byte numbers, 146
Single chip PSP, 244
Software development, 36
  development aids, 195
  development process, 147
  for the PSP, 246
  multiplication, 267
Solar activity, 299
Source program, 202
Stack, 121
  pointer, 111, 121
Starting address, 28
Static RAM, 95
Status information, 155
  latches, 168
  lines, 12
Storage CRO, 284
Subroutine, 142
  jumps, 121
Successive approximation method, 257
  register, 258
System clock module, 155
  configuration, 22
  control, 297
  controller, 154
  design, 34
  programs, 202
  restart program, 180
  unix, 203

Target machine, 196

Teletypewriter terminal, 15
Terminals, 16
Testing, 38
Time-frequency transformation, 223
Time-shared computer systems, 25
Transducer, 21
Transform, 217
  domain, 217
  Fourier, 99, 201
  Haar, 230
  Walsh, 224
Transistor switching, 11
Transistor-transistor logic (TTL), 71
Transportation of data matrices, 232
Traps, 32
Truth table, 43
Turnkey, 199
Two-pass assembly, 136
Typewriter golfball, 16

UCSD Pascal system, 149, 199
  screen editor, 203
Unconditional branch, 118
Unix system, 203
Up-counter, 87
Up–down counter, 88
Utilities, 32

Vector, 146
  interrupt system, 180
Versatile interface adapter, 303
Very large scale integration, 215
Virtual memory, 30
  address, 28, 107
Visual display unit (VDU), 17
Voltage comparator, 251

Walsh function, 230
  transform, 230
Washing machine control, 5
Western Digital, 305
Wordlength, 9
Word-parallel systems, 217
WRITE operation, 10

Zilog Z-80 Microprocessor, 104
  Z8000 Microprocessor, 30, 150, 166